Ergonomics for Beginners

What the

"   ld be on
t   *Back*

"

onomics

This re   *ers* provides a   ing the
subject   to the
perform   mbrac-
ing the   human
comfort   ce, the
book sh   ion, the
steps b   appropriate h

To   with
a focu

Jan D   ntract
Resea   enced
resear   eerd-
meest   le, a
privat

Ergonomics for Beginners
A quick reference guide

Second edition

Jan Dul and Bernard Weerdmeester

Taylor & Francis
Taylor & Francis Group

Boca Raton   London   New York   Singapore

A CRC title, part of the Taylor & Francis imprint, a member of the
Taylor & Francis Group, the academic division of T&F Informa plc.

First published 1993 by Taylor & Francis
11 New Fetter Lane, London EC4P 4EE

Second edition 2001

Simultaneously published in the USA and Canada
by Taylor & Francis Inc,
29 West 35th Street, New York, NY 10001

Reprinted 2003

Taylor & Francis is an imprint of the Taylor & Francis Group

© 2001 Jan Dul and Bernard Weerdmeester

This English language edition published by kind permission of
Kluwer Bedrijfswetenschappen. Based on a translation by R.E. Vande
Putte from *Vademecum Ergonomie*, originally published in Dutch in
1963. © 1991 Kluwer Bedrijfswetenschappen.

Typeset in Sabon by Wearset, Boldon, Tyne and Wear
Printed and bound in Great Britain by TJ International Ltd, Padstow,
Cornwall

Every effort has been made to ensure that the advice and information
in this book is true and accurate at the time of going to press.
However, neither the publisher nor the authors can accept any legal
responsibility or liability for any errors or omissions that may be
made. In the case of drug administration, any medical procedure or
the use of technical equipment mentioned within this book, you are
strongly advised to consult the manufacturer's guidelines.

British Library Cataloguing in Publication Data
A catalogue record for this book is available from the British Library

Library of Congress Cataloging in Publication Data
Dul, Jan, 1954-
    Ergonomics for beginners : a quick reference guide / Jan Dul and
Bernard Weerdmeester.
        p. cm.
    Includes bibliographical references and index.
    1. Human engineering. I. Weerdmeester, B.A. (Bernard A.) 1955-
II. Title.

TA 166 .D78 2001
620.8'2–dc21   00-053265

ISBN 0-7484-0825-8 (pbk. : alk. paper)

Contents

Preface

As generations succeed each other, people's expectations change, building on historical experiences so that what was accepted by one generation becomes unacceptable to those which follow. What was at one time a relatively local phenomenon, with modern communication has become world-wide; living and working conditions are subject to common demands. The European Community is a clear example of this trend, both for reasons of social justice and of economy this common market requires that working conditions across its member states shall be broadly equivalent.

In pursuit of this equivalence, basic regulatory measures are passed, and many of these now call, overtly, for ergonomic solutions to work problems. By introducing ergonomics, the clear intent is that the old style of work design, where the operator was viewed as a 'pair of hands' is not acceptable. People are, in the terms of ergonomics, to be seen in the round, as complete people making a contribution to their work on a more human level than as 'hewers of wood and drawers of water'.

It is thus very appropriate that this new and revised edition of an early Dutch classic, the *Vademecum ergonomie* should be published at this time. Its increased breadth of coverage and updated content provide a comprehensive summary of what is important in the application of ergonomics to the world of work. Both as an aid to the implementation of ergonomics and as a ready source of reference it will contribute to the improvement of the workplace, not just in the West, but in any industrial enterprise.

Of course there is no pretence by the authors to imply that this is a sufficient manual for all ergonomics problems. But the wider understanding of ergonomics and how to apply it that this book can bring about will increase the recognition of what the subject can do. The benefits which workpeople, union representatives, industrial engineers and management can themselves achieve by this book will deepen their understanding and make it more possible to ease the path of ergonomics specialists who deal with the more intractable problems.

This is a book to be welcomed for the 21st century, embracing as it does the concept of designing work for human satisfaction as well as human feasibility. Advanced industrial societies have shown the gains to be made from companies which run as a partnership rather than as a battleground. This little book shows the steps by which management and workforce can advance towards that better state.

E.N. Corlett

Foreword

The original *Vademecum ergonomie* by Kellermann, Klinkhamer, Van Wely and Willems was first published in Dutch in 1963. Its success was such that within five years it had been translated into eleven languages. The pace of developments in ergonomics over the past few years has created an increasing need for a fundamental revision of this guide. In 1991 we presented the first English edition of the book which was a complete revision of the earlier editions, but nevertheless retains the basic approach of the original guide. We are glad that the revised edition was also so well received that it has become a best seller.

This is not a book to be read once only, but rather it is meant to be a reference text which covers the subject through easily understood practical guidelines and advice. Each recommendation is summarized in one line printed in **sans serif bold** in the centre of the page, followed by a short explanation. This means that the user can either rapidly become acquainted with ergonomics, or can quickly refresh his or her knowledge in certain areas, or can broaden such knowledge.

The text of *Ergonomics for Beginners* is pitched at the introductory level, covering the subject in six chapters. The first chapter is an introduction describing ergonomics and explaining its social significance. Chapters 2, 3, 4 and 5 provide basic ergonomics knowledge about posture and movement, information and operation, environmental factors as well as work organization. In Chapter 6 we present an ergonomic approach which can be used in the design or acquisition of production systems, machines, accessories, consumer products and suchlike. In this approach, basic ergonomic knowledge is applied as a whole. The guide concludes with a chapter which provides information to those who wish to learn more about the subject.

In this second edition of *Ergonomics for Beginners* we have updated several chapters. Due to the rapid developments in information and communication technologies and thinking on management and organization, Chapter 3 on 'Information and Operation' and Chapter 5 on 'Work Organization' were completely revised. Chapter 7 on 'Sources of Additional Information' now contains also a list of ergonomics websites.

Jan Dul
(dul@ergonomicsforbeginners.com)

Bernard Weerdmeester
(weerdmeester@ergonomicsforbeginners.com)

1 Introduction

Ergonomics developed into a recognized field during the Second World War, when for the first time, technology and the human sciences were systematically applied in a co-ordinated manner. Physiologists, psychologists, anthropologists, medical doctors, work scientists and engineers together addressed the problems arising from the operation of complex military equipment. The results of this inter-disciplinary approach appeared so promising that the co-operation was pursued after the war, in industry. Interest in the approach grew rapidly, especially in Europe and the United States, leading to the foundation in England of the first ever national ergonomics society in 1949, which is when the term 'ergonomics' was adopted. This was followed in 1961 by the creation of the International Ergonomics Association (IEA), which at present represents ergonomics societies which are active in 40 countries or regions, with a total membership of some 15 000 people.

What is ergonomics?

The word 'ergonomics' is derived from the Greek words 'ergon' (work) and 'nomos' (law). In the United States, the term 'human factors' is often used. A succinct definition would be that ergonomics aims to design appliances, technical systems and tasks in such a way as to improve human safety, health, comfort and performance. The formal definition of ergonomics, approved by the IEA, reads as follows:

> Ergonomics (or human factors) is the scientific discipline concerned with understanding of the interactions among humans and other elements of a system, and the profession that applies theory, principles, data and methods to design, in order to optimize human well-being and overall system performance.

In the design of work and everyday-life situations, the focus of ergonomics is man. Unsafe, unhealthy, uncomfortable or inefficient situations at work or in everyday life are avoided by taking account of the physical and psychological capabilities and limitations of humans. A large number of factors play a role in ergonomics; these include body posture and movement (sitting, standing, lifting, pulling and pushing), environmental factors (noise, vibration, illumination, climate, chemical substances), information and operation (information gained visually or through other senses, controls, relation between displays and control), as well as work organization (appropriate tasks, interesting jobs). These factors determine to a large extent safety, health, comfort and efficient performance at work and in everyday life. Ergonomics draws its knowledge from various fields in the human sciences and technology, including anthropometrics, biomechanics, physiology, psychology, toxicology, mechanical engineering, industrial design, information technology and industrial management. It has gathered, selected and integrated relevant knowledge from these fields. In applying this knowledge, specific methods and techniques are used. Ergonomics differs from other fields by its interdisciplinary approach and applied nature. The interdisciplinary character of the ergonomic approach means that it relates to many different human facets. As a consequence of its applied nature, the ergonomic approach results in the adaptation of the workplace or environment to fit people, rather than the other way round.

What is an ergonomist?

In some countries it is possible to graduate as an ergonomist. Other people who are trained in one of the relevant basic technical, medical or social science fields can also acquire knowledge of, and capabilities in, ergonomics through training and experience. In several countries, professional ergonomists can be certified by independent certifying bodies. In Europe, for example the Center for Registration of European Ergonomists (CREE) decides on candidates for registration as European Ergonomists (EurErg). Professional ergonomists can work for the authorities (legislation), training institutions (universities and colleges), research establishments, the service industry (consultancy) and the production sector (occupational health services, personnel departments, design departments, research departments, etc.).

Many professional ergonomists who are active in business (company ergonomists) practise their profession mainly by being an intermediary between, on the one hand, the designers and, on the other hand, the users of production systems. The ergonomist highlights the areas where ergonomic knowledge is essential, provides ergonomic guidelines and advises designers, purchasers, management and employees, on which are the more acceptable systems. There are other experts besides professional ergonomists who make use of ergonomic knowledge, methods and techniques. These would include, for example, industrial designers, company doctors, company nurses, physiotherapists, industrial hygienists, and industrial psychologists.

Social significance of ergonomics

Ergonomics can contribute to the solution of a large number of social problems related to safety, health, comfort and efficiency. Daily occurrences such as accidents at work, in traffic and at home, as well as disasters involving cranes, aeroplanes and nuclear power stations can often be attributed to human error. From the analysis of these failures it appears that the cause is often a poor and inadequate relationship between operators and their task. The probability of accidents can be reduced by taking better account of human capabilities and limitations when designing work and everyday-life environments.

Many work and everyday-life situations are hazardous to health. In western countries diseases of the musculoskeletal system (mainly lower back pain) and psychological illnesses (for example, due to stress) constitute the most important cause of absence due to illness, and of occupational disability. These conditions can be partly ascribed to poor design of equipment, technical systems and tasks. Here, too, ergonomics can help reduce the problems by improving the working conditions. Therefore, in a number of countries, occupational health services are obliged to employ ergonomists.

Finally, ergonomics can contribute to the prevention of inconveniences and also, to some considerable degree, can help improve performance. In the design of complex technical systems such as process installations, (nuclear) power stations and aircraft, ergonomics has become one of the most important design factors in reducing operator error. Some ergonomic knowledge has been compiled into official standards whose objective is to stimulate the application of ergonomics. A range of ergonomic subjects is covered by international ISO standards of the International Standardization Organization (ISO), European EN-standards of the Comité Européen de Normalisation (CEN), as well as national standards, for example in the United States (ANSI) and Britain (BSI). In addition, there are specific ergonomic standards which are applied in individual companies and in industrial sectors.

General and individual ergonomics

An important ergonomic principle is that equipment, technical systems and tasks have to be designed in such a way that they are suited to every user. The variability within populations is such that most designs, in the first instance, are suited to only 95 per cent of the population. This means that the design is less than optimum for five per cent of the users, who then require special, individual ergonomic measures. Examples of groups of users, who from an ergonomic perspective require additional attention, are short or tall persons, overweight people, the handicapped, the old, the young, and pregnant women.

Ergonomics for Beginners focuses primarily on the application of ergonomics in a more general sense. Individual requirements for special groups cannot be dealt with in a book of this size.

2 Posture and movement

Posture and movement play a central role in ergonomics. At work and in everyday life, postures and movements are often imposed by the task and the workplace. The body's muscles, ligaments and joints are involved in adopting a posture, carrying out a movement and applying a force. The muscles provide the force necessary to adopt a posture or make a movement. The ligaments, on the other hand, have an auxiliary function, while the joints allow the relative movement of the various parts of the body. Poor posture and movement can lead to local mechanical stress on the muscles, ligaments and joints, resulting in complaints of the neck, back, shoulder, wrist and other parts of the musculoskeletal system. Some movements not only produce a local mechanical stress on the muscles and joints, but also require an expenditure of energy on the part of the muscles, heart and lungs. In the following sections we shall begin by providing some general background on posture and movement. Thereafter, possibilities for optimizing tasks and the workplace are presented for commonplace postures and movements such as sitting, standing, lifting, pulling and pushing.

Biomechanical, physiological and anthropometric background

A number of principles of importance to the ergonomics of posture and movement derive from a range of specialist fields, namely biomechanics, physiology and anthropometrics. These general principles are discussed in this section and are applied in the subsequent sections (see p. 12 and p. 28) to some specific postures and movements.

Biomechanical background

In biomechanics, the physical laws of mechanics are applied to the human body. It is thereby possible to estimate the local mechanical stress on muscles and joints which occurs while adopting a posture or making a movement. A few biomechanical principles of importance to the ergonomics of posture and movement are outlined below.

Joints must be in a neutral position

When maintaining a posture or making a movement, the joints ought to be kept as far as possible in a neutral position. In the neutral position the muscles and ligaments which span the joints are stretched to the least possible extent, and are thus subject to less stress. In addition, the muscles are able to deliver their greatest force when the joints are in the neutral position.

Raised arms, bent wrists, bent neck and turned head, bent and twisted trunk are examples of poor postures where the joints are not in a neutral position.

Keep the work close to the body

If the work is too far from the body, the arms will be outstretched and the trunk bent over forwards. The weight of the arms, head, trunk and possibly the weight of any load being held then exerts a greater horizontal leverage on the joints under stress (elbow, shoulder, back). This obviously increases the stress on these muscles and joints. Figure 2.1 shows that the stress to the back increases when the arms are outstretched.

Avoid bending forward

The upper part of the body of an adult weighs about 40 kg on average. The further the trunk is bent forwards, the harder it is for the muscles and ligaments of the back to maintain the upper body in balance. The stress is particularly large in the lower back. Prolonged bending over for long periods must therefore be avoided wherever possible.

A twisted trunk strains the back

Twisted postures of the trunk cause undesirable stress to the spine. The elastic discs between the vertebrae are stretched, and the joints

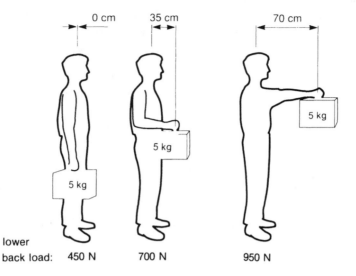

Figure 2.1 Increasing the distance between the hands and the body increases the stress on, among others, the lower back (10 N is about 1 kg force).

and muscles on both sides of the spine are subjected to asymmetric stress.

Sudden movements and forces produce peak stresses

Sudden movements and forces can produce large, short-duration stresses. These peak stresses are a consequence of the acceleration in the movement. It is well known that sudden lifting can cause acute back pain in the lower back. Lifting must occur as far as possible in an even and gradual manner. Thorough preparation is necessary before large forces are exerted.

Alternate postures as well as movements

No posture or movements should be maintained for a long period of time. Prolonged postures and repetitive movements are tiring, and in the long-run can lead to injuries to the muscles and joints. Although the ill-effects of prolonged postures and repetitive movements can be prevented by alternating tasks, it is best to avoid movements which involve regular lifting or repetitive arm movements. Likewise, standing, sitting and walking should also be alternated and it should be possible to carry out prolonged tasks either standing or sitting.

Limit the duration of any continuous muscular effort

Continuous stress on certain muscles in the body as a result of a prolonged posture or repetitive movement leads to localized muscle fatigue, a state of muscle discomfort and reduced muscle performance. As a result, the posture or movement cannot be maintained continuously. The greater the muscular effort (exerted force as a percentage of the maximum force), the shorter the time it can be maintained (Figure 2.2).

Most people can maintain a maximum muscular effort for no more than a few seconds and a 50 per cent muscular effort for no more than approximately two minutes as this causes muscular exhaustion.

Prevent muscular exhaustion

The muscles will take a fairly long time to recover if they become exhausted which is why exhaustion must be avoided. Figure 2.3

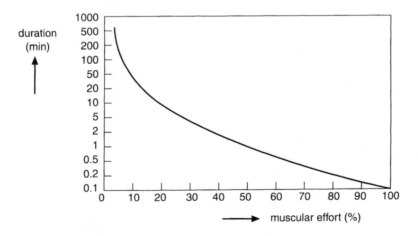

Figure 2.2 The duration of continuous localized muscular effort must be limited. The figure shows the relationship between muscular effort (exerted force as a percentage of maximum force) and the maximum possible duration (in minutes) of any continuous muscular effort.

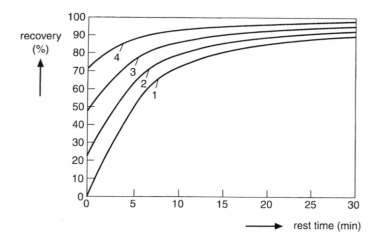

Figure 2.3 Recovery curves for muscles which have been exhausted (curve 1) or partially exhausted (curves 2 to 4) after continuous muscular effort.

shows an example of recovery curves after a muscle has been partially or totally exhausted from continuous effort.

In this example, an exhausted muscle needs to rest for 30 minutes to achieve a 90 per cent recovery. Muscles in a half-exhausted state will recover to the same degree after 15 minutes. Complete recovery can take many hours.

More frequent short breaks are better than a single long one

Muscular fatigue can be reduced by distributing the resting time over the task duration or working day. It is not sensible to accumulate break times until the end of the task or working day.

Physiological background

In exercise physiology, estimates are made of the energy demands on the heart and lungs resulting from muscular effort during movements. In addition to fatigue which results from continuous localized muscular effort (see Biomechanical background, p. 6), general body fatigue can develop from carrying out physical tasks over a long period. The limiting factor here is the amount of energy which the heart and lungs can supply to the muscles to allow postures to be adopted or movements to be carried out. A few physiological principles of importance to the ergonomics of posture and movement are discussed below.

Limit the energy expenditure in a task

The majority of the population can carry out a prolonged task without experiencing any general fatigue provided the energy demand of the task (expressed as the energy consumed by the person per unit of time) does not exceed 250 W (1 W = 0.06 kJ min^{-1} = 0.0143 kcal min^{-1}). This figure includes the amount of energy, approximately 80 W, which the body needs when at rest. At the energy consumption level given above, the task is not considered heavy, and no special measures such as breaks, or alternation with light activities, are necessary for recovery. Examples of activities with an energy demand of less than 250 W are writing, typing, ironing, assembling light materials, operating machinery, a gentle walk or leisurely cycle ride.

Rest is necessary after heavy tasks

If the energy demand during a task exceeds 250 W, then additional rest is necessary to recover. Rest can be in the form of breaks or less demanding tasks. The reduction in activity must be such that the average energy demand over the working day does not exceed 250 W.

Table 2.1 lists some activities with a high energy demand. It is also true here that rest is most effective if the total rest time is spread over a number of break periods spaced regularly during the task, and not saved up until the end of the task or the end of the working day.

Table 2.1 Examples of activities with an energy demand in excess of 250 W. Additional measures are necessary to avoid exhaustion in the long term (breaks, alternation with lighter activities, etc.)

Activity	Energy expenditure
Walking while carrying a load (30 kg, 4 km hr^{-1})	370 W
Frequent lifting (1 kg, 1 × per sec)	600 W
Running (10 km hr^{-1})	670 W
Cycling (20 km hr^{-1})	670 W
Climbing stairs (30 deg, 1 km hr^{-1})	960 W

Anthropometric background

Anthropometry is concerned with the size and proportions of the human body. A few anthropometric principles of importance to the ergonomics of posture and movements are given below.

Take account of differences in body size

The designers of workplaces, accessories and suchlike must bear in mind differences in body size of the potential users. A table height which is suitable for a person of average stature can be unsuitable for a tall or short person. A table height which is adjustable over a sufficient range is the solution if the table is to be used by several people.

Sometimes only the shortest users must be considered, for example, in designing a control panel which has to be reached with the arms. In other cases, such as in choosing a door height, the tall users have to be considered instead.

Use the anthropometric tables appropriate for specific populations

Data for body dimensions always refer to a particular population group and do not necessarily apply to other population groups. Table 2.2, for example, shows the body dimensions of British adults. The adult population of Great Britain is relatively tall in comparison with the average world population. The dimensions refer to unclothed, unshod persons. Some 3–5 cm must be added to the stature to account for shoe thickness. The data in the table do not apply to other population groups. To gain an idea of the extent of individual variation in body size, the data in Table 2.2 are given for:

- short adults (only 5 per cent of adult females are shorter)
- the average person
- tall persons (only 5 per cent of adult males are taller).

The average height of a British adult is 1.68 m, that of a short British adult is 1.51 m or less, and that of a tall British adult is over 1.85 m. The correlation between body dimensions in Table 2.2 is limited. For instance, a person with a short lower arm (dimension 15 is small) could have a long trunk (dimension 12 is large).

Table 2.2 Body sizes of short, average and tall British adults. All measurements are in centimetres, except for body weights, which are in kilograms

	short	average	tall
Standing			
1. Stature	150.5	167.5	185.5
2. Forward grip reach	65.0	74.3	83.5
3. Chest depth	21.0	25.0	28.5
4. Vertical grip reach	179.0	198.3	219.0
5. Eye height	140.5	156.8	174.5
6. Shoulder height	121.5	136.8	153.5
7. Elbow height	93.0	104.8	118.0
8. Knuckle height	66.0	73.8	82.5
Sitting			
11. Sitting height	79.5	88.0	96.5
12. Sitting eye height	68.5	76.5	84.5
13. Sitting elbow height	18.5	24.0	29.5
14. Popliteal height	35.5	42.0	49.0
15. Elbow-grip length	30.4	34.3	38.7
16. Buttock-popliteal length	43.5	48.8	55.0
17. Buttock-knee length	52.0	58.3	64.5
00. Body weight	44.1	68.5	93.7

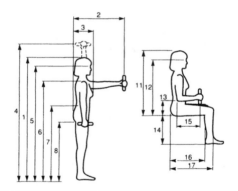

Posture

Posture is often imposed by the task or the workplace. Prolonged postures can in time lead to complaints of the muscles and joints. In this section we take a look at the stress due to prolonged sitting and standing, as well as that due to hand and arm postures, such as would occur in the use of hand-held tools.

Select a basic posture that fits the job

The characteristic of the job determines the best basic posture: sitting, standing p. 20, combinations of sitting and standing (sit-stand work stations p. 23) or work stations with pedestal stools p. 24. Figure 2.4 provides a selection procedure for the best basic posture.

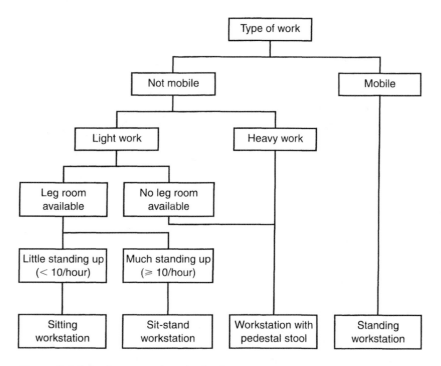

Figure 2.4 Selection procedure for basic posture.

Sitting

Working for long periods in a seated position occurs mostly in offices, but also occurs in industry (assembly and packaging work, sometimes for machine operation). Sitting has a number of advantages compared to standing. The body is better supported because several support surfaces can be used: floor, seat, back rest, armrest, work surface. Therefore, adopting this posture is less tiring than standing. However, activities which require the operator to exert a lot of force or to move around frequently are best carried out standing (see p. 20).

Alternate sitting with standing and walking

Although sitting is usually more favourable than standing, sitting for long periods should be avoided as it has a number of disadvantages. Many manual activities carried out while seated (e.g. writing or assembly work) require the person to keep the hands in view. This means that the head and trunk have to be bent forward. The neck and back are then subjected to prolonged stress which can lead to neck and back complaints. Bending the trunk forward also means that the back rest of the chair is no longer used. The back is subject to further stresses if the trunk has to be twisted and the seat cannot swivel. Manual work often requires working with unsupported raised arms, which can lead to shoulder complaints.

Tasks which require prolonged sitting (for example, at a VDU screen) should be alternated with tasks which can be carried out in a standing position, or where walking is necessary. A sit-stand work-place (p. 23), or a chair that promotes active sitting allow the user to alternate between sitting and other postures during the task.

The height of the seat and back rest of the chair must be adjustable

There are many ergonomically-sound chairs on the market. The most important general feature of such chairs is that the height of the seat and back rest is adjustable.

- It must be possible to adjust the height of the seat while sitting, in a continuous, smooth motion rather than in steps. For British adults, the minimum adjustment range should be at least 13 cm between the heights of 39 and 52 cm, based on popliteal height differences (measure 14 in Table 2.2 plus 3 cm for shoe thickness).
- The height of the seat must be chosen in such a way that when the feet are properly supported, the upper legs are also properly supported, without the back of the knee being cramped.
- The backrest must provide support mainly to the lower back (for British adults, the minimum adjustment range should be at least 10 cm between the heights of 20 and 30 cm, based on differences in the lumbar height, not shown in Table 2.2). Avoid misusing a low back rest as a high back rest.
- The lower part of the back rest must be given a convex shape in order to preserve the curve of the lower back.

In addition, the chair should swivel. This reduces the need to twist the body.

Limit the number of adjustment possibilities

Adjustment possibilities must be restricted to only the most important components of the chair; as a minimum, the seat height and the height of the back rest. If too many features are adjustable, settings will be used either incorrectly or not at all.

Provide proper seating instructions

Users of adjustable chairs must receive regular instruction in the optimum adjustment of the chair, say every six months. This also applies to other adjustable elements of the workplace, such as the table.

Specific chair characteristics are determined by the task

In addition to its general characteristics, an ergonomically-sound chair will also display specific features which depend on the task. A chair with armrests can be selected if these do not hinder the activities, as armrests can partly support the weight of the arms and trunk, and are also useful when rising from the chair. Armrests should be short to allow close proximity to the table. Castors can be useful if a chair has to be moved frequently but none should be present if pedals have to be operated. If the trunk is mostly upright or tilted somewhat backwards, the seat ought to be tilted backwards a few degrees. For tasks where the body is unavoidably bent forward, a limited forward tilt (maximum 20 degrees), is advantageous, as it can prevent the lower back from curving. Figure 2.5 shows an example of an ergonomically-sound chair for VDU work. Here, the seat and back rest are adjustable, the back rest supports also the lower back, and the short armrests and castors provide additional comfort.

The work height depends on the task

The chair is only one of several factors determining whether the working posture is correct. The position of the hands as well as the focal point are also of great importance to the posture of the head, trunk and arms. The correct height for the hands and focal point

Figure 2.5 An ergonomically-sound chair for VDU work. The height of the
seat and back rest (with support for the lower back) can easily be
adjusted. The chair swivels, has short, adjustable armrests and is
fitted with castors.

depends on the task, individual body dimensions and individual preference. During most tasks the hands have to be used and viewed
simultaneously. Then, the work height is a compromise between the
optimum height for the arms and the optimum position of the head
and trunk. In the first instance, a low table is better since the arms
have to be raised to a lesser extent and it is easier to apply a force. In
the second instance, a high table is better because it means less
bending forward and a better view of the work.

General guidelines for the work height are given in Table 2.3 for
three types of tasks. These guidelines apply both when sitting or
standing.

The height of the hands and focal point need not always be the
same as the table or work surface height. The work surface might
have to be lowered, to take into account the thickness of the workpieces, tools or accessories (e.g. a keyboard). The work surface and

Table 2.3 Guidelines for the height of the hands and focal point, for carrying out various tasks while seated or standing

Type of task	Work height
Use of eyes: frequent, Use of hands/arms: infrequent	10–30 cm below eye height
Use of eyes: frequent, Use of hands/arms: frequent	0–15 cm above elbow height
Use of eyes: infrequent, Use of hands/arms: frequent	0–30 cm below elbow height

the objects on it should not be too thick otherwise legroom will be restricted.

Work tables used for a given type of seated task not involving objects of different thickness must be adjustable over a range of at least 25 cm because of differences between individuals. Where a number of tasks have to be carried out which require different work heights, the adjustment range must be even greater.

A good starting position for the height of a VDU workstation is one where the hands are kept at elbow height. The height of a VDU table with a keyboard thickness of 3 cm (measured at the position of the middle row of keys) must be adjustable between 54 and 79 cm for British adults (based on differences in sitting elbow height, see measures 13 and 14 in Table 2.2). It must be possible to make the adjustment easily from the seated position.

The heights of the work surface, seat and feet must be compatible

In a seated workplace which can be adjusted individually, the vertical distances between the feet, seat and working surface must be compatible. The height of the feet is mostly fixed because they rest on the floor. The chair and table must then be adjustable and set according to the guidelines of Table 2.3.

Use a footrest if the work height is fixed

If the work height cannot be adjusted by the individual user, such as at a machine, a relatively high work surface must be chosen to suit tall users. The seat height is then adjusted to the work surface. The height of the feet should then also be adjusted, using a suitable footrest, which would be not simply a bar, but a slightly sloping surface.

Avoid excessive reaches

It is necessary to limit the extent of forward and sideways reaches to avoid having to bend over or twist the trunk. Workpieces, tools and controls which are in regular use should be located directly in front of, or near the body.

Figure 2.6 shows the reach envelopes in three planes. The most important operations should take place within a radius of approximately 50 cm. This value applies to both seated and standing work. Application of the guidelines on reaches is given in Chapter 3, which covers the design of control panels.

Select a sloping work surface for reading tasks

If the activities allow it, use should be made of a sloping work surface for reading tasks and other tasks where the work has to be kept in view, such as writing and assembly work without tools. A sloping work surface brings the work to the eye instead of the other

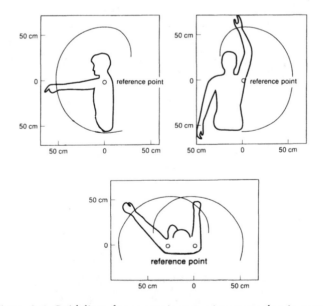

Figure 2.6 Guidelines for convenient maximum reaches in seated or standing work.

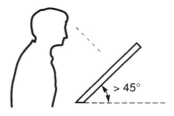

Figure 2.7 In tasks which require no manual work, such as reading, bending the head and trunk forward can be reduced by using a sloping work surface of at least 45 degrees for viewing.

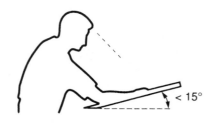

Figure 2.8 Viewing manual work.

way round, thereby improving the posture of the head and trunk. Because the height of the front of the table or machine remains the same, the arms do not have to be raised any further. Successful use of a sloping work surface requires measures to prevent workpieces or accessories from sliding off (non-slip work surface, rim, etc.) or alternatively, only part of the work surface should slope. Sloping work surfaces can often be created easily by raising the back of the table or machine or by using a lectern. For reading purposes, the position of the work surface which is viewed, must be tilted by at least 45 degrees (Figure 2.7).

For tasks where the hands have to be used and kept in view, such as in writing, the work surface must be placed at an angle of approximately 15 degrees (Figure 2.8). A greater slope is not desirable because of insufficient support for the arms and because objects may slide off.

Allow sufficient legroom

Sufficient legroom must be provided under the work surface (Figure 2.9). The width clearance must be at least 60 cm. The required depth

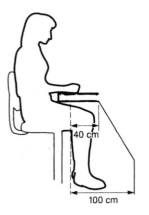

Figure 2.9 Required legroom for sitting.

clearance must be at least 40 cm at the knees and 100 cm at the feet, which should allow the user to sit close to the work without bending the trunk forward.

It is desirable to be able to stretch the legs once in a while when sitting for long periods. To this effect the depth clearance at the feet should be at least 1 m.

To have sufficient room between the underside of the working surface and the top of the legs, the thickness of the working surface (and objects on it) must be as small as possible. The thickness of a writing surface, for example, should not exceed 3 cm.

Standing

Activities where considerable force has to be exerted or where the work place has to be frequently changed, should be carried out in a standing position.

Alternate standing with sitting and walking

It is not recommended for the whole working day to be spent in a standing position. Standing for long periods tires the back and legs. An additional stress can arise when the head and trunk are bent, leading to neck and back complaints. Furthermore, working with the arms unsupported, in a raised position, leads to stress on the shoulders, which may result in shoulder complaints.

Tasks which have to be carried out over long periods in a standing position should be alternated with tasks which can be carried out while seated, or with tasks where walking is required. People should also be given the opportunity to sit down during natural breaks in the work (e.g. in the case of operating a machine or in sales work in shops). A sit-stand workplace or a pedestal stool will allow the user to vary postures during the task (see pp. 23 and 24).

The work height depends on the task

The work height for standing work depends, as for seated work, on the task, on individual preference, and on individual body dimensions. Table 2.3 contains the guidelines for the optimum work height for different types of standing tasks carried out at a work surface.

The height of the work table must be adjustable

It must be possible to adjust the height of a work table which is intended for use by several people (as a result of part-time working, team work or task rotation), or whenever different tasks (e.g. with varying sizes of workpieces) must be carried out at the same table. It must be possible to conveniently adjust the table from the normal working position. A table meant for standing work which is used for a given task, and on which no objects of different thickness are used, must have an adjustment range of at least 25 cm in order to cater for individual differences in body size. Users must be instructed in the optimum height of the table.

Do not use platforms

The use of platforms for standing work is not advisable. The major disadvantages of platforms are that they constitute a trip hazard, are cumbersome to clean, and hamper transport along floors. They also require additional work space, and are not practical if their height has to be regularly adjusted for different people or to different working heights.

Provide sufficient room for the legs and feet

Sufficient room must also be kept free under the work surface or machine for the legs and feet in standing work. This allows the person to be close to the work without bending the trunk. Enough clearance is also required for changing the position of the legs once in

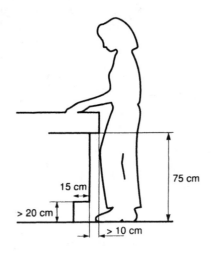

Figure 2.10 Minimum foot and leg-room required for standing work
(dimensions in cm).

a while. Figure 2.10 illustrates the required minimum recesses under
the work surface or machine.

Avoid excessive reaches

Forward and sideways reaches must be limited in order to avoid
having to bend forward or twist the trunk. Workpieces, tools and
controls which are in regular use should be located directly in front
of, and near, the body. Convenient maximum reaches are given in
Figure 2.6.

Select a sloping work surface for reading tasks

If the activities allow it, a sloping work surface should be used for
reading tasks just as in the case of seated work. The same is true also
for other tasks where the work must be kept in view, such as writing.
Guidelines for a sloping work surface are given in Figures 2.7 and
2.8.

Change of posture

This section describes ways of relieving prolonged postures. These
techniques relate to the provision of a varied task package, the

implementation of a sit-stand workplace and the use of a pedestal stool.

Offer variation in tasks and activities

The design and organization of activities should ensure that everyone is given variation in tasks and activities so that no prolonged postures occur. The principle of job enrichment can be usefully applied (see Chapter 5).

Introduce sit-stand work stations

If tasks have to be carried out over a long period, the workplace should be adapted to allow the work to be carried out either standing or sitting. To this effect a work height is selected which is suitable for standing work (see Standing, p. 20). In addition, a special high chair also allows the work to be carried out while seated. Leg-room has been left under the work surface and a footrest is provided (Figure 2.11).

Alternate sitting postures

A prolonged seated posture can be varied by using different types of chairs. There are chairs available that promote 'active sitting'. The chairs offer possibilities to change posture and have adjustable seats

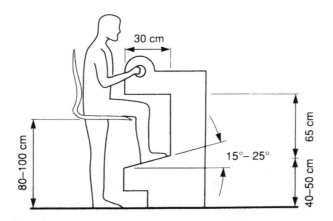

Figure 2.11 Guidelines for the dimensions of a workplace at which seated and standing work can be alternated.

and backs, allowing movements of the body. Despite the use of these chairs, it is still advisable to alternate sitting with standing and walking.

Make occasional use of a pedestal stool in standing work

A pedestal stool can be used once in a while to vary a standing work posture. A pedestal stool consists of a seat which is adjustable in height (65–85 cm), and is tilted forwards between 15 and 30 degrees. It allows semi-supported postures to be adopted, which somewhat relieve the stress on the legs. A pedestal stool cannot be used for long periods and is only suited to standing activities where large forces or extensive movements are not required. The floor on which it rests must provide sufficient friction to prevent the support from sliding away (Figure 2.12).

Hand and arm postures

Working for long periods with the hand and arm in a poor posture can lead to specific complaints of the wrist, elbow and shoulder. A continuously bent wrist can lead to local nerves becoming inflamed and trapped, resulting in wrist pain and a tingling sensation in the fingers. Another ailment is tennis elbow, which is a local inflamma-

Figure 2.12 A pedestal stool can be used to change posture when standing for long periods.

tion of a tendon attachment due to a combination of a bent elbow and bent wrist.

Neck and shoulder complaints occur in prolonged work with unsupported, raised arms. These problems arise especially from handling tools. In addition to posture, application of a force and repetitive movement ('Repetitive Strain Injuries or RSI') play a role in the development and aggravation of these conditions. Correct hand and arm postures can be promoted by selecting a correct working height for the hands (see Table 2.3) and by using the right tools (see below).

Select the right model of tool

A particular tool is often available in different models. Select a model which is best suited to the task and posture, so that the joints can be kept as far as possible in the neutral position. Figure 2.13 shows the correct and incorrect use of different types of electric drills and screwdrivers.

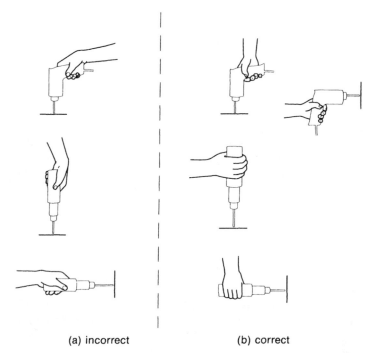

(a) incorrect   (b) correct

Figure 2.13 When using hand-held tools, the wrist should be kept as straight as possible. The figure shows the correct and incorrect use of two types of rotating tool.

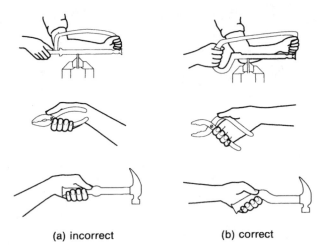

(a) incorrect   (b) correct

Figure 2.14 Correct location of handgrips on tools avoids having to bend
the wrist.

Do not bend the wrist, use curved tools instead

Bending the wrist can be prevented by correctly locating the handgrips on a tool (Figure 2.14).

Hand-held tools must not be too heavy

If the tool cannot rest on a surface, and is normally used with one
hand, its weight should not exceed 2 kg. If the tool can rest on a
surface, heavier weights are allowed, but the maximum weights that
apply to lifting (p. 30) must be taken into consideration.

Heavy tools which are frequently used can be suspended on a
counterweight (Figure 2.15).

Maintain your tools

Proper maintenance of tools can contribute to a reduction in bodily
stress. Blunt knives, saws or other equipment require greater force.
Proper maintenance of motorized hand-held tools can also reduce
wear, noise and vibration (see Chapter 4).

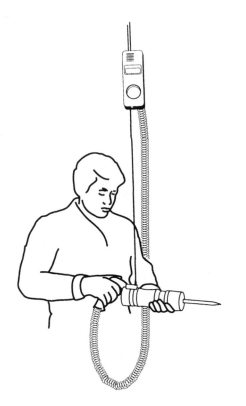

Figure 2.15 The weight of heavy hand-held tools can be supported by a counterweight.

Pay attention to the shape of handgrips

The shape and location of handgrips on trolleys, loads, machines, equipment and suchlike must take into consideration the position of the hands and arms. If the whole hand is used to exert a force, the handgrip must have a diameter of approximately 3 cm and a length of approximately 10 cm (Figure 2.16).

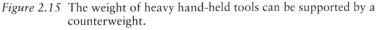

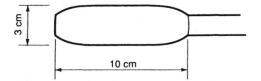

Figure 2.16 Shape of handgrips on hand-held tools.

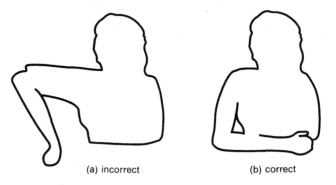

(a) incorrect   (b) correct

Figure 2.17 Hand and elbow positions above shoulder height are to be
avoided.

The handgrip must be somewhat convex to increase the contact
surface with the hand. The use of pre-shaped handgrips is not
advised: the fingers are constrained, too little account is taken of
individual differences in finger thickness and the grips are not suitable for use with gloves.

Avoid carrying out tasks above shoulder level

The hands and elbows should be well below shoulder level when carrying out a task (Figure 2.17). If work above shoulder level is
unavoidable, the duration of the work must be limited and regular
breaks must be taken.

Avoid working with the hands behind the body

Working with the hands behind the body should be avoided (Figure
2.18). This kind of posture occurs when sliding away objects, for
example, at check-outs in supermarkets.

Movement

Various tasks require moving the whole body, often while exerting a
force. Such movements can cause high, localized mechanical stresses
which in time can lead to bodily aches and pains. Movements can
also be stressful in the energetic sense for the muscles, heart and
lungs. In this section we examine the stress from lifting, carrying,
pulling and pushing.

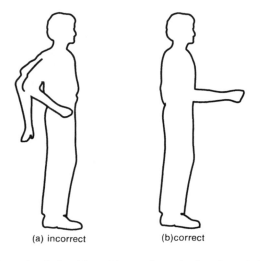

(a) incorrect   (b)correct

Figure 2.18 Avoid positions where the hands and elbows are held behind the body.

Lifting

Manual lifting is still frequently needed in spite of mechanization and automation. Lifting is seen as a major cause of lower back complaints. Many lifting situations still do not satisfy ergonomic requirements. This section contains guidelines and measures for lifting. These measures relate to the production technique (mechanical or manual), the work organization (task design, lifting frequency), the workplace (position of load with respect to the body), the load (shape, weight, presence of handgrips), lifting accessories, and the working method (lifting by several persons, individual lifting technique).

Restrict the number of tasks which require displacing loads manually

Production systems must be designed to use mechanization as a way of restricting the extent of manual lifting. In this case, however, attention must be paid to new problems relating to posture and movement. These could include prolonged manual operation of machines or lifting accessories, or the necessity for heavy maintenance work on machines which are difficult to access. Other problems can also develop as a result of mechanization, for example, noise and vibration, monotony, and reduced social contacts.

If it is not possible to avoid heavy or frequent lifting, these activities must be alternated with other (light) activities, for example, by applying job enrichment (see Chapter 5).

In lifting, but also in other physical activities, it is important that the work pace should be set by the person involved. It is essential to avoid situations where the rate of lifting is imposed by a machine, by colleagues or by a supervisor.

Create optimum circumstances for lifting

If manual lifting of heavy loads (up to 23 kg) is necessary, then lifting conditions have to be optimized:

- it must be possible to hold the load close to the body (horizontal distance from hand to ankles about 25 cm);
- the initial height of the load before it is lifted should be about 75 cm;
- the vertical displacement of the load should not exceed 25 cm;
- it must be possible to pick up the load with both hands;
- the load must be fitted with handles or hand-hold cut-outs;
- it must be possible to choose the lifting posture freely;
- the trunk should not be twisted when lifting;
- the lifting frequency should be less than one lift per five minutes;
- the lifting task should not last more than one hour, and should be followed by a resting time (or light activity) of 120 per cent of the duration of the lifting task.

Ensure that people always lift less, and preferably much less, than 23 kg

Only in the above-mentioned optimum lifting situation may a person lift the maximum load of 23 kg. Lifting conditions, however, are virtually never optimum, in which case the maximum allowable load is considerably less (see below). The load should not exceed a few kilogrammes if it has to be picked up far away in front of the body and has to be displaced over a large vertical distance.

Use the NIOSH method to assess lifting situations

In practice, optimum lifting conditions are seldom met, and therefore the permitted load is much less than 23 kg. The method developed by the American National Institute for Occupational Safety and Health

(NIOSH) can be used to determine the maximum load in unfavourable lifting conditions. This takes into account the horizontal and vertical distance between the load and the body, the trunk rotation, the vertical displacement, the lifting frequency and duration, and the coupling between hands and load. It assumes among other things that the lifting posture can be freely chosen and that the load is lifted with both hands. The NIOSH equation is devised in such a way that the weight is acceptable for the majority of the population (99 per cent of men and 75 per cent of women) that the compressive load on the lower back is less than 3400 N (340 kg force), and that the energy expenditure for 1–2 hours repetitive lifting is less than 260 W for lifts below the bench height (75 cm) and less than 190 W for lifts above the bench height. In the NIOSH method, the unit weight of 23 kg is reduced for unfavourable lifting conditions by using a series of multipliers according to the formula:

Recommended weight limit
$$= 23\,\text{kg} \times HM \times VM \times DM \times FM \times AM \times CM$$

Figure 2.19 shows the multipliers for the horizontal load distance (horizontal multiplier, HM), the vertical load distance (vertical multiplier, VM), the vertical displacement of the load (displacement multiplier, DM), the frequency (frequency multiplier, FM), the asymmetric factor (asymmetric multiplier, AM), and the coupling factor (coupling multiplier, CM). If the lifting situation does not satisfy the requirements of the NIOSH method (e.g. if the lifting posture cannot be freely chosen, or if the load is lifted with one hand), the method will result in values that are too high. Because of the complexity of the analysis, several software packages have been developed to analyse lifting situations including combinations of different lifting tasks, using the NIOSH method. These can also help to develop improvements based on the results of the analysis.

Individual loads should not be too light

The weight of a load (e.g. the unit weight of a packaging) has to be chosen carefully. On the one hand the NIOSH recommendation should not be exceeded under normal conditions. On the other hand the loads should not be too light otherwise more frequent lifting becomes necessary. If individual loads are too light there is also a danger that several loads may be lifted simultaneously.

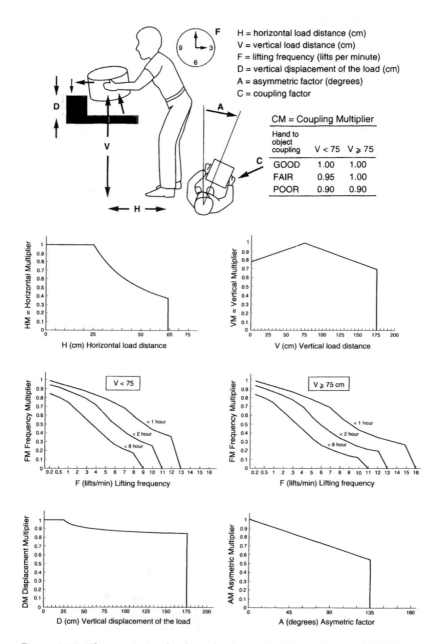

Figure 2.19 The maximum load can be determined for unfavourable lifting conditions by using the NIOSH equation.

Make the workplace suitable for lifting activities

The design of tables, shelves, machines and suchlike, onto which loads have to be placed or from which loads have to be lifted, must result in optimum lifting conditions being achieved.

- It must be possible to approach the load properly when lifting and setting it down.
- Foot and legroom must be sufficient to allow a stable position for the feet and to be able to bend the knees.
- Twisting the trunk should not be necessary.
- The height and location of the load on the work surface must be such that when lifting the load or setting it down, the hands are at the optimum height of approximately 75 cm, and close to the trunk.

It can generally be said that the measures described here have considerable influence on the admissibility of lifting activities. It is often possible to achieve more through a higher and closer position of the load than by reducing its weight.

Loads should be fitted with handgrips

A load should be fitted with two handgrips so that it can be grasped with both hands and lifted (Figure 2.20b). Grasping the load with the fingers (Figure 2.20a) should be avoided because far less force can be exerted.

The position of the handgrips should be such that the load cannot twist when lifted.

(a) incorrect   (b) correct

Figure 2.20 No handgrips (a) and correct (b) handgrips for lifting loads.

Ensure that the load is of the correct shape

The size of the load must be as small as possible so that it can be held close to the body. It must be possible to move the load between the knees if it has to be lifted from the floor. The load should not have any sharp edges nor be hot or cold to the touch. For special loads, such as a container of hazardous liquid, or a hospital patient, additional attention should be paid to the lifting process, for instance, by taking special safety precautions and planning the lifting operation.

If a person is to lift loads of a weight unknown in advance to him or her, it is desirable to label the loads in advance showing their weight and possibly advise caution.

Use correct lifting techniques

Sometimes a person can more or less freely choose the lifting technique. In such instances, prior training will ensure that the best possible posture is adopted during the lift. On the other hand, the benefits of information and training should not be overestimated. In practice, improved lifting techniques are often not feasible because of restrictions at the workplace. In addition, ingrained habits and movements can only be changed after intensive training and repetition.

Training should address the following aspects:

• assess the load and establish where it must be moved to. Consider using the help of others or the use of a lifting accessory;
• where lifting has to be done without any additional help, stand directly in front of the load. Make sure the feet are in a stable position. Bring the load as close as possible to the body. Grasp the load with both hands, using the whole hand, not just a few fingers;
• hold the load as close as possible to the body while lifting. Make a flowing movement with a straight trunk. Avoid twisting the trunk. If necessary, move the feet.

The latter recommendation is of great importance in reducing back stress (Figures 2.21 and 2.22). Bending or twisting of the trunk while lifting, contribute significantly to injuries of the lower back. The load in Figure 2.21 is approximately 20 kg; bending the trunk forward, as in (a) results in a back stress nearly 30 per cent greater than in (b).

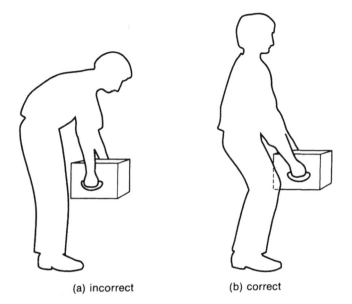

(a) incorrect　　　　　　　(b) correct

Figure 2.21 Lifting with a bent trunk and a large horizontal distance between load and lower back (a) is more hazardous than lifting with the back straight and a small horizontal distance between load and body (b).

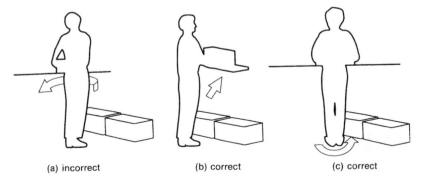

(a) incorrect　　　　(b) correct　　　　(c) correct

Figure 2.22 Twisting the trunk while lifting (a) must be avoided by a better choice of lay-down surface (b), or by moving the feet (c).

Heavy lifting should be done by several people

Several people can work together if the load is too heavy to be lifted by one person. The partners must be of approximately the same height and strength, and must be able to work well together. One of them must co-ordinate the lifting as this will prevent unexpected movements.

Use lifting accessories

Many lifting accessories are available to help lift and move loads. The different types include, for example, levers, raising platforms, and cranes. Figure 2.23 gives an example of a special device for lifting kerbstones (a), a dedicated mobile lift for moving patients (b), and two universal accessories: a mobile lifting table (c) and a crane (d).

Carrying

After a load has been lifted it must sometimes be moved manually. In general, walking with a load is both mechanically stressful and

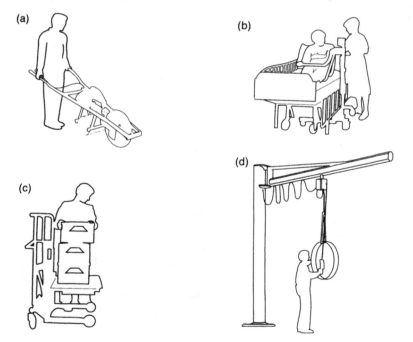

(a)
(b)
(c)
(d)

Figure 2.23 Examples of lifting accessories.

energetically demanding. As a result of holding the load, the muscles are subjected to continuous mechanical stress; this particularly affects the muscles in the arms and back. Displacing the whole body and the load consumes energy.

Limit the weight of the load

The permissible weight of a carried load is determined mostly by the lifting which precedes the carrying. Guidelines for lifting are given in the section on lifting (p. 29).

Hold the load as close to the body as possible

To limit both mechanical stress and energy consumption, the load must be kept as close as possible to the body. Small, compact loads are therefore preferable to larger loads. By using accessories such as a backpack or a yoke, it is possible to hold the load even closer to the body.

Provide well-designed handgrips

The load should be fitted with well-designed handgrips that have no sharp edges. Alternatively, an accessory such as a hook may be used instead.

Avoid carrying tall loads

A person lifting a tall load will tend to bend the arms to prevent the load from hitting his or her legs. This causes additional fatigue to the muscles in the arms, shoulders and back. The vertical dimension of the load must therefore be limited (Figure 2.24).

(a) incorrect   (b) correct

Figure 2.24 Carrying of tall loads should be avoided.

Avoid carrying loads with one hand

When only one hand is used to carry a load, the body is subject to an asymmetric stress; well-known examples of this are carrying a school bag, suitcase or shopping bag. The solution is to carry two lighter loads (one in each hand) or use a backpack.

Use transport accessories

There is a large number of provisions and accessories such as roller conveyors, conveyor belts, trolleys and mobile raising platforms which make it unnecessary to carry loads manually. Whenever one of these is selected, the user must be aware of any possible new problems resulting from the lifting, pulling and pushing required to place the load on the device or move it (trolley, etc.).

Pulling and pushing

Many types of trolley have to be moved manually. Pulling and pushing trolleys places stress mainly on the arms, shoulders and back. The design of the trolley must take this into account (Figure 2.25).

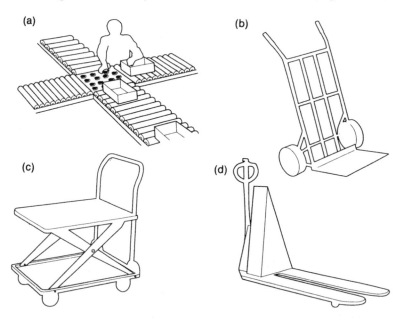

Figure 2.25 Transport accessories which replace manual carrying: (a) roller conveyor, (b) sack barrow, (c) mobile raising platform, (d) forklift.

Limit the pulling and pushing force

When setting a trolley in motion by pulling or pushing, the exerted manual force should not exceed approximately 200 N (about 20 kg force). Although the maximum possible force required is often considerably higher, this limit should be adhered to in order to prevent large mechanical stresses, mainly to the back. If the trolley is kept moving for more than one minute, the permissible pulling or pushing force drops to 100 N.

In practice this means that trolleys with a total weight (including load) of 700 kg or more should certainly not be displaced manually. The permissible weight depends on the type of trolley, the kind of floor, the wheels, and so forth. Many types of motorized trolley are available which can be used as an alternative.

Use the body weight when pulling or pushing

A correct pulling and pushing posture is one which uses the body's own weight. When pushing, the body should be bent forwards and when pulling, it should lean backwards. The friction between the floor and the shoes must be sufficiently large to allow this. There must also be sufficient clearance for the legs to be able to maintain this posture. In pulling and in pushing, the horizontal distance between the rearmost ankle and the hands must be at least 120 cm. When pulling, there must also be room under the trolley to place the forward foot directly below the hands (Figure 2.26).

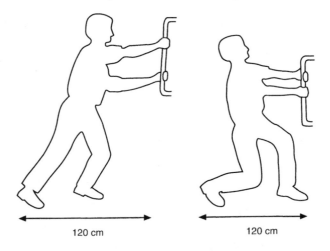

120 cm   120 cm

Figure 2.26 Using the weight of the body when pushing or pulling trolleys.

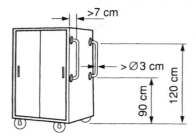

Figure 2.27 Recommended handgrip designs for pulling and pushing trolleys.

Provide handgrips on trolleys

Trolleys and suchlike should be fitted with handgrips so that both hands can be fully utilized to exert a force. The dimensions (in cm) of handgrips for pushing and pulling are given in Figure 2.27. The handgrips must be cylindrical. Vertical handgrips, at a height of 90–120 cm, have the advantage that the hands can be placed at the right height when maintaining a correct pulling or pushing posture.

A trolley should have two swivel wheels

Trolleys used on a hard floor surface must be fitted with large, hard wheels, which are able to limit any resistance due to unevenness of the floor. Two swivel wheels should be fitted to achieve good manoevrability. These should be positioned on the side that is pushed or pulled, i.e. where the handgrips are located. Having four swivel wheels is not advisable because this makes it necessary to steer continuously.

The loaded trolley must not be higher than 130 cm so that most persons can see over the load while pulling or pushing.

Ensure that the floors are hard and even

If possible, avoid having to lift trolleys over any raised features such as kerbs. If this is unavoidable, the trolleys should be fitted with horizontal handgrips. The weight to be lifted may not exceed the limits given above for lifting (p. 31).

SUMMARY CHECKLIST

Biomechanical, physiological and anthropometric background

Biomechanical background

1 Are the joints in a neutral position?
2 Is the work held close to the body?
3 Are forward-bending postures avoided?
4 Are twisted trunk postures avoided?
5 Are sudden movements and forces avoided?
6 Is there a variation in postures and movements?
7 Is the duration of any continuous muscular effort limited?
8 Is muscle exhaustion avoided?
9 Are breaks sufficiently short to allow them to be spread over the duration of the task?

Physiological background

10 Is the energy consumption for each task limited?
11 Is rest taken after heavy work?

Anthropometric background

12 Has account been taken of differences in body size?
13 Have the right anthropometric tables been used for specific populations?

Posture

14 Has a basic posture been selected that fits the job?

Sitting

15 Is sitting alternated with standing and walking?
16 Are the height of the seat and back rest of the chair adjustable?
17 Is the number of adjustment possibilities limited?
18 Have good seating instructions been provided?
19 Are the specific chair characteristics dependent on the task?
20 Is the work height dependent on the task?
21 Do the heights of the work surface, seat and feet correspond?
22 Is a footrest used where the work height is fixed?

23 Are excessive reaches avoided?
24 Is there a sloping work surface for reading tasks?
25 Is there enough legroom?

Standing

26 Is standing alternated with sitting and walking?
27 Is work height dependent on the task?
28 Is the height of the work table adjustable?
29 Has the use of platforms been avoided?
30 Is there enough room for the legs and feet?
31 Are excessive reaches avoided?
32 Is there a sloping work surface for reading tasks?

Change of posture

33 Has an effort been made to provide a varied task package?
34 Have combined sit-stand workplaces been introduced?
35 Are sitting postures alternated?
36 Is a pedestal stool used once in a while in standing work?

Hand and arm postures

37 Has the right model of tool been chosen?
38 Is the tool curved instead of the wrist being bent?
39 Are hand-held tools not too heavy?
40 Are tools well maintained?
41 Has attention been paid to the shape of handgrips?
42 Has work above shoulder level been avoided?
43 Has work with the hands behind the body been avoided?

Movement

Lifting

44 Have tasks involving manual displacement of loads been limited?
45 Have optimum lifting conditions been achieved?
46 Has care been taken that any one person always lifts less, and preferably much less, than 23 kg?
47 Have lifting situations been assessed using the NIOSH method?
48 Are the weights to be lifted not too light?
49 Are the workplaces suited to lifting activities?

50 Are handgrips fitted to the loads to be lifted?
51 Does the load have a favourable shape?
52 Have good lifting techniques been used?
53 Is more than one person involved in heavy lifting?
54 Are lifting accessories used?

Carrying

55 Is the weight of the load limited?
56 Is the load held as close to the body as possible?
57 Are good handgrips fitted?
58 Is the vertical dimension of the load limited?
59 Is carrying with one hand avoided?
60 Are transport accessories being used?

Pulling and pushing

61 Are pulling and pushing forces limited?
62 Is the body weight used during pulling and pushing?
63 Are the trolleys fitted with handgrips?
64 Do the trolleys have two swivel wheels?
65 Are the floors hardened and even?

3   Information and operation

Increasing numbers of people are making use of complex products and systems. This leads to interaction based on receiving information, and acting on it. The relationship between information and operation, and the elements which play a role in this, can be represented by a man–machine model (Figure 3.1).

Since computers play a major role in daily life, this chapter will concentrate on working with computers. Most of the principles which are discussed here are not new and certainly do not apply exclusively to working with computers; applications from other fields are given in the examples.

The boundary between man and machine is called the user-interface. This chapter will deal with optimization of the user-interface by means of better presentation of information, easy controllers and a good interaction between the presented information and the operation (called the dialogue).

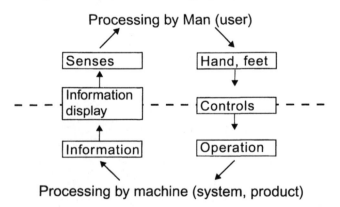

Figure 3.1 A man–machine model shows the relationship between information and operation.

The user

When designing computer programs (or information systems) it is important to know who the users of the program will be. The possibilities and limitations of these persons determine to a large extent how the user-interface should be designed in an optimal way. The user's limitations are a particularly critical factor, because users have to absorb more and more information at higher speeds and complexity.

Determine the user-population as accurately as possible

When determining the user-population, the following properties are important:

• nationality (language and culture) and age;
• knowledge and level of education;
• available controls;
• task to perform (what is the program used for?);
• frequency of use;
• possibilities for education and training.

Take mental models into account

In the memory of the user an image of the program to be operated is formed. This is called a mental model: an impression of the program with all the relevant aspects to facilitate imitating the behaviour of the program. Mental models have a role in the use of the program and also in interpreting the behaviour of it. Some examples of mental models are:

• Maps: in their memory people have an image of the pattern of the streets of a city to find their way around.
• Analogies of metaphors: by comparing a system with another system the operation becomes clear; for instance a computer-screen with objects from an office environment (the desk-top metaphor of Apple and Windows)
• Organization schema: by means of a diagram of an organization, people are able to position others within the organization.

Mental models are not the same for all users. Especially when using metaphors, it is very important to investigate if the chosen metaphor is familiar to the users.

Information

Visual information

Simultaneous perception of a large amount of information by humans is best achieved through the eyes. This makes the eyes the most important source of information, and means that people with only limited eyesight will miss much information, or will only assimilate it slowly. The form in which the information is presented must be suited to as many people as possible. Perception of information by means of sound and other senses will be mentioned shortly.

Characters

Ergonomics literature provides guidelines on the legibility of screens, books, newspapers and magazines, some of which are reproduced here. The guidelines are also valid for information-transfer, for instance by means of overhead sheets, computer presentations, slides and commercial posters.

Do not use text consisting entirely of capitals

In continuous text, lower-case letters are preferable to upper-case letters. The letters with ascenders (b, d, f, h, k, l, t) and those with descenders (g, j, p, q, y) stand out and contribute to the image of a word (Figure 3.2). The reader can see at a glance what is written and need not read letter by letter. Capitals can be used for the first letter in a sentence, for a title or proper noun and for abbreviations that are familiar to the user.

Do not justify text by inserting blank spaces

Right-justifying text by inserting blank spaces does not contribute to good legibility. Proportional writing, with less space for narrow letters such as 'i' and more space for wide letters such as 'm' and 'w', is more pleasant to read, but is not essential.

KEYHOLE   Keyhole

(a) incorrect   (b) correct

Figure 3.2 Text consisting entirely of capital letters is not as legible.

Serif letters   Sanserif letters
(a) incorrect   (b) correct

Figure 3.3 Characters without much ornamentation are the most legible.

Use a familiar typeface

Plain characters without much ornamentation are without doubt the most legible. In particular headings and captions (directions, names, book titles) a sanserif typeface is preferable to a serif typeface (Figure 3.3).

Avoid confusion between characters

Some characters are difficult to distinguish from each other, which can lead to confusion. The smaller the number of points forming the character, the greater the risk of confusion. In a normal text this will not lead to many difficulties, but in abbreviations and for letters, numbers and code numbers, this could well be confusing. Some frequent causes of confusion are shown in Figure 3.4.

Make sure that the characters are properly sized

The required dimensions of characters depend on the reading distance. A rule of thumb is that the height of capital letters should be at least $\frac{1}{200}$th of the reading distance. Letters presented in a conference room 20 metres long should be at least 10 centimetres high on the screen. On VDU screens, capitals should be no smaller than 3 mm. The requirements relating to proportions are given in Figure 3.5. It is sensible in the case of height and width of capitals to base

Mutual	One-way
O and Q	C mistaken for G
T and Y	D mistaken for B
S and 5	H mistaken for M or N
I and L	J,T mistaken for I
X and K	K mistaken for R
I and 1	2 mistaken for Z
O and 0 (zero)	B mistaken for R,S or 8

Figure 3.4 Avoid confusion between similar characters.

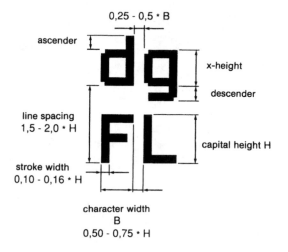

Figure 3.5 Use characters of the correct shape.

these on the capital 'O' and for lower case letters, to base these on the lower case 'o'. The stroke width for letters in a text should preferably be based on a non-round letter, for example the capital 'I'.

The longer the line, the greater the required line spacing

In a text, the required distance between lines, that is, the distance between the imaginary lines on which the letters are placed, depends on the length of the line. The line spacing in newspaper columns can therefore be much less than in a book (Figure 3.6). A guideline here is that the line spacing should be at least $\frac{1}{30}$th of the line length. If the lines are too closely spaced, the eye finds it difficult to follow on from the end of one line to the beginning of the next.

Good contrast contributes to legibility

Whether characters are legible or not depends on contrast, that is, the difference in brightness between the text and the background. Contrast has an even greater influence on legibility than lighting. Some examples are illustrated in Figure 3.7.

　　For VDUs, other technical characteristics of the screen such as flicker frequency will also play a role. Within a given task, it is not desirable to change frequently from dark symbols on a light back-

In a text, the required distance between lines, that is, the distance between the imaginary lines on which the letters are placed, depends on the length of the line. The line spacing in newspaper columns can therefore be much less than in a book (Figure 3.7). A guideline here is that the line spacing should be a least $\frac{1}{30}$th of the line length. If the lines are too closely spaced, the eye finds it difficult to follow on from the end of one line to the beginning of the next.

In a text, the required distance between lines, that is, the distance between the imaginary lines on which the letters are placed, depends on the length of the line. The line spacing in newspaper columns can therefore be much less than in a book (Figure 3.7). A guideline here is that the line spacing should be a least $\frac{1}{30}$th of the line length. If the lines are too closely spaced, the eye finds it difficult to follow on from the end of one line to the beginning of the next.

Figure 3.6 Longer lines require a wider line spacing.

ground to light symbols on a dark background. For word processing from a draft on paper it is therefore preferable, for example, to use a white screen with black letters. This requires the refreshing rate of the screen to be fairly high to prevent flickering.

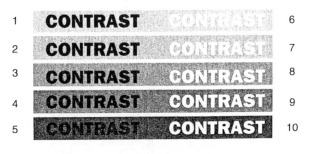

1 and 10 are the most legible.

Figure 3.7 Choose a good contrast.

Diagrams

Diagrams should be used to support text or as a substitute for text. The use of diagrams has become even more attractive now that computer programs are available which can help those who are not skilled enough to draw them. However, the number of options offered by the computer is, in fact, so large that users with little graphical knowledge may well produce diagrams which are incomprehensible.

Produce diagrams that are easy to understand

Diagrams should be such that they can be understood by everyone, and captions should be legible. Legibility is especially important for presentations with beamers, overhead or slide projectors. The correct letter size will enhance legibility (see page 47). This often requires reducing the amount of text, and therefore the user is encouraged to dispense with all irrelevant information. Here are a few guidelines:

• titles and captions should not consist entirely of capitals;
• abbreviations must make sense; do not allow the computer to truncate words at random;
• different types of shading must be easy to distinguish from each other;
• optical vibration caused by moire patterns must be avoided;
• the scale should match the units being used (e.g. 0–100 for percentages);
• the subdivision of the scale must be sensible.

Use pictograms with care

In principle, symbols are not bound to a particular language. It seems a good idea to replace written signs by symbols such as pictograms or icons, in public places where many people from different language groups come together. Many of these pictograms are, in actual fact, poorly understood. Here are a few guidelines:

• bear in mind cultural differences: a picture of a knife and fork can mean a simple snack bar in one country or a luxury restaurant in another;
• people should have a simple mental image of what is meant to be represented by the pictogram; in many people's minds a door can

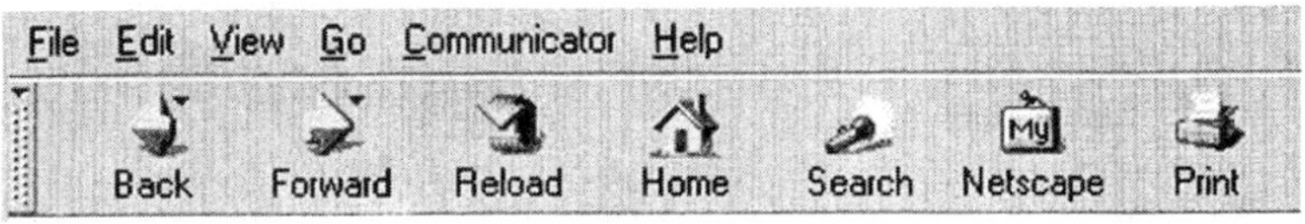

Figure 3.8 Icons for a popular Internet browser.

represent both an entrance or an exit: it is therefore difficult to represent these concepts by two different pictograms;

* more stylized pictograms, i.e. further removed from reality, are less well understood;
* use only pictograms which represent a single concept, not a combination of concepts.

In an Internet browser, a pocket torch is used to search for information, but this icon has no relation to the application. The icon for reloading the Internet page is also not easy to understand without written explanation (Figure 3.8).

Perception of visual information

Technical developments are making it increasingly easy to collect and store data. Selecting the appropriate data from this multitude and interpreting it become all the more difficult. A few general principles can help make this easier.

Select the appropriate method for displaying information

The choice of display is very closely linked to the objective. Two types of display are shown in Figure 3.9. Here are a few guidelines:

* pointers are suited for global estimation and for the perception of rapid changes. In this case the pointer itself must move and the complete dial must be shown;
* exact information is best represented by numbers;
* recording instruments are suitable for representing slow variations and historical reviews.

Type of meter	 ⌐⌐⌐ /	 2 1 5 4 5
Reading a value	–	+
Estimating a value, controlling a state	+	+
Setting to a required value	–	+
Controlling an on-going process	+	o

+ = good   o = average   – = poor

Figure 3.9 Adapt the display to the objective.

Hearing

The ear is not often used in information perception, except in communication through speech. Nevertheless, if the eyes are overtaxed in a particular task it may sometimes be possible to rely on the ears instead. Frequent application of auditory signals is not recommended, however, as even the most pleasant sounds can start to irritate in the long run. The use of speech seems attractive. It is possible to imitate speech by a computer in such a quality that a user does not notice the difference in comparison to a real voice. Sometimes this difference is essential: it determines the way the user reacts. An example is computer-speech via the telephone (voice-response systems).

Sound should be reserved for warning signals

The ear is particularly suited to detecting warning signals, because it is difficult for people to escape from sounds that come from all directions. In contrast to a warning given by a light signal, a person does not have to be at a particular spot to detect a warning sound. If a task consists of noticing when limit values are exceeded, the number of missed signals will increase with time if vision is the only sense used, but the task is easier if a bell or buzzer sounds when the limit is exceeded.

Select the correct pitch

If the distance between the device and the operator is large, the sound intensity must be large and the pitch not too high. Low pitches are more suitable if the sound has to travel around corners and obstacles.

The signal must be clearly distinguishable from background noises, as is the case, for example, with sirens on fire engines and police cars. A high-pitched signal is better if the background includes many low-pitched sounds. The signal must be regularly interrupted where there is continuous background noise.

Synthesized speech must have adjustable features

It is easier for a computer to synthesize speech than to recognize speech. The user should be able to control computer speech in the following way:

- alter its speed (faster or slower);
- produce a repeat of any section;
- interrupt (pause);
- switch over to another form of display.

Attempts to make computer speech sound human are not always desirable. In many instances the user wants to be able to hear the difference between a computer and a person.

Other senses

Humans possess three other senses apart from sight and hearing: smell, taste and touch. These senses can also be used as sources of information.

Restrict the use of taste, smell and temperature to warning signals

Taste, smell and temperature should only be used to indicate alarm conditions; an example is the addition of odour to natural gas. These senses should be used only sparingly. Furthermore, they are inadequate for multiple use. For example, two smells present together cannot easily be distinguished from each other.

Use the sense of touch for feedback from controls

The sense of touch can be used to provide feedback on the location and status of controls. It is particularly useful in places outside the visual field. Examples of this are the bass buttons on an accordion or the identifying marks on the '5' of a numerical keypad (Figure 3.10).

Use different senses for simultaneous information

Modern information technology allows the simultaneous presentation of a large amount of information. If the information comes through different senses, this multiplicity will present no problem: driving a car while the radio is on, is no more difficult than with the radio off, but reading a book when observing a control panel is more difficult because the same sense organs are being used together. On the other hand, if variations or alarm states are conveyed by a warning bleep, reading can indeed take place while controlling. It is recommended that important information such as alarms be conveyed via several senses simultaneously. A good example is a sound, coupled with a light signal. An example where many senses are simultaneously used is the use of telephones while driving cars: the amount of information to digest and the load of the memory are both very high.

Controls

People transmit their ideas or decisions to machines by means of controls, such as knobs, handles or steering wheels. Computer systems, on the other hand, are controlled mostly via keyboards. In this chapter we discuss how controls operate.

A separate, but important, point to bear in mind is that the forces which are necessary to actuate the controls also need to be

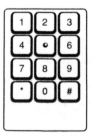

Figure 3.10 Make positions on a keyboard perceptible to touch.

considered. Indeed, these forces should be as small as possible, but not so small that unintentional operation can easily occur.

Distinguishing between controls

Controls must be given easily distinguishable shapes whenever the use of the wrong control could have disastrous consequences.

Make controls distinguishable by touch

The simplest way to indicate differences between controls is to vary their shape and size, with differences in size of at least 30 per cent (Figure 3.11). An example of a failure to do this is when coins of different value are about the same size. Sometimes it is also possible to make their actual effect distinguishable by touch, as in an 'on–off' light switch, for example.

Use a standard location and provide sufficient spacing

Confusion can arise if controls used in similar circumstances are located in different positions, or if the spacing between the controls is insufficient. This is particularly important for controls which have an opposite effect, such as the accelerator and brake in a car.

Avoid unintentional operation

Touching a control unintentionally, for example with the hand or arm, could change its state, without being noticed. In some instances, the consequences could be disastrous. Unintentional operation can be avoided by adjusting the mode of operation (rotating knob instead of

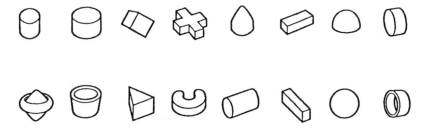

Figure 3.11 Controls should have different shapes, to make them distinguishable from each other.

push button) or the location of the control (countersunk into the keyboard). The risk of unintentional operation exists with:

- handles that can be moved in different ways, each having a different effect (e.g., a combination of indicator and light switch in a car);
- keys which do not have to be pressed in but only touched, so-called fingertip controls. The risk of confusion is much greater if the controls are closely spaced. In many cases these should be replaced by so-called membrane keys which have to be pressed in to some degree. The use of fingertip controls and membrane keys must be limited to locations where dirt could impair the proper operation of mechanical keys (controls in digital metal-working machines) or where there are strict hygiene requirements (food industry, medical equipment).

Controls should be placed well within reach

Controls should not be located too far away from the operator, and should not obstruct other controls. The following considerations apply in the location of controls:

- the importance of the control;
- the order in which the control operations are carried out;
- the frequency with which the control is used.

Controls that have the highest priority must be placed in an optimum location. The others should be within the 'maximum reach envelope' (Figure 3.12).

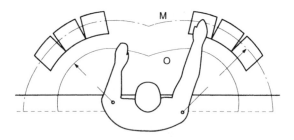

M = maximum reach
O = optimum reach

Figure 3.12 The design of the workstation should take account of the importance, order and frequency of use of controls.

Think carefully before using labels and symbols

The use of labels or symbols on controls may seem a good idea. The number of possibilities is considerable, but the prerequisites for use are:

- sufficient space;
- sufficient lighting;
- adequate letter size;
- location close to or on the control;
- use of straightforward, everyday words;
- use of readily understood symbols;
- information being restricted to the meaning of the control.

A familiar example of the use of labels is on a keyboard. Although mostly capital letters appear on the keys, lower case letters appear on the screen.

Limit the use of colour

Although the eye can distinguish between a large number of colours, it is advisable to use only the following five colours for colour discrimination of controls: red, orange, yellow, green and blue. Five points should be kept in mind:

- the difference with respect to the background colour, and the contrast;
- the association which people make with some colours (red for danger, green for safe);
- reduced colour discrimination (colour blindness);
- the colour and lighting of the surroundings;
- colour strongly attracts attention; its use should therefore be limited.

Types of controls

Nowadays information transfer often takes place by means of a keyboard or a mouse. The touch screen is sometimes used (at public terminals or information pillars). The focus of the information on the screen is often indicated by means of a small block, line or arrow (the cursor). There are forms of controls for instance, for hands-free use of remote controls. Even the old handwheels, knobs and handles have not disappeared totally.

Use the QWERTY keyboard layout

The QWERTY keyboard layout has been in use for a long time. In this layout, the top row of letters starts with the letters Q-W-E-R-T-Y (Figure 3.13). The most important advantage of this layout is its worldwide distribution and acceptance as a standard. Placing the letters in alphabetical order, or other types of arrangement, certainly does not make a system quicker to use.

Select a logical layout for the numerical keypad

There are three international standards for the layout of a numerical keypad: the standard for pocket calculators with 1-2-3 on the bottom row, the standard for push-button telephones with 1-2-3 on the top row, and a layout where all keys are in a line, such as in the top row of keys on a keyboard (Figure 3.14). The choice of the most suitable layout for a particular task should take into consideration which of the three layouts is the most logical or least confusing.

Figure 3.13 QWERTY is a useful layout for keyboards.

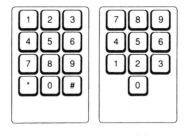

Figure 3.14 Select a standard layout for numerical keypads, one which does not cause confusion.

Restrict the number of function keys

Function keys can be used to avoid typing system commands letter by letter. However, the meaning of a function key must be clear under all circumstances. It is advisable to distinguish function keys from other keys through size, shape, colour or position. Function keys of similar shape are best grouped in sets of three or four, because it is quite easy to select the outer key in a row or the one next to it. There are three types of function keys:

- hard (independent) function keys, which have a fixed meaning, irrespective of application. Familiar examples are the correction keys and the cursor keys. They can also be used for the following purposes: start/stop (on/off), help, do (execute), restore, delete, and print;
- programmable function keys, where the user determines the meaning. This could be, say, a frequently used sequence of system commands;
- variable functions keys, where the meaning depends on the current application and is determined by the software. In this case, it is sensible for the meaning to appear at a corresponding position (Figure 3.15).

Match the type of cursor control to the task

The best method of pointing out something on a screen in a given situation depends, among other things, on the required speed, accuracy and direction of movement. Input can be with a joystick, the

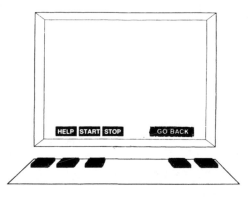

Figure 3.15 When using softkeys, the meaning of the function keys must be clear.

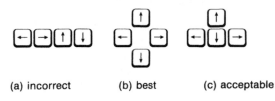

(a) incorrect   (b) best   (c) acceptable

Figure 3.16 The relative position of cursor keys must correspond to the direction of movement.

user's finger, a light pen, functions keys or a mouse. If we are dealing with text, cursor keys are adequate for indicating position, but their relative location on the keyboard must correspond to the direction of movement (Figure 3.16).

Do not use the mouse exclusively

Frequent use of the mouse can cause an overload and one-sided load of certain muscles and tendons in hands and wrists ('Repetitive Strain Injuries', RSI). It is therefore recommended to alternate the use of the mouse with the use of the keyboard, for instance by using so-called 'short-cuts'. These are sometimes represented as combinations of keys using the 'ALT'-key, but also typing the first (underlined) letter of a menu item is a short-cut.

Touch-screens are suitable for inexperienced users

Touch-screens offer a suitable form of dialogue with a computer if inexperienced, casual users need to retrieve information from data-bases (Figure 3.17). An advantage of this method is the direct relationship between what the eyes see and what the hands do. The disadvantages are that:

- the arms are often in an uncomfortable, stretched position;
- the areas to be selected on the screen need to be relatively large;
- the screen must be close to the user;
- the screen becomes dirty.

Use pedals only if the use of the hands is inconvenient

Pedals are an option whenever a large force is to be exerted, or if both hands are otherwise needed for precise control. The number of

Figure 3.17 Touch-screens will help inexperienced users in searching.

pedals in any situation should not exceed three. The position of both feet should not differ too much while operating the pedals. Frequent use of pedals should be avoided in activities that are carried out while standing.

Remote controls give the user more freedom

The use of remote controls has advantages if the user cannot be, or does not wish to be, in direct contact with the system being controlled. Think, for example, of industrial cranes, or of home video recorders. Here are a few guidelines:

- the design must be such that the direction of operation is evident (see Figure 3.18);
- if several systems have to be operated remotely from a single point, this must be done by using only one remote control. No more than two appliances or systems should be controlled remotely at any other time;
- important function keys (e.g. on/off) must have a fixed location, for example in the upper right-hand corner. The meaning must be consistent (always 'out' or 'stand-by').

Relationship between information and operation

Operations of systems takes place in the form of interaction: after an action of the user the system reacts, which results in a reaction of the

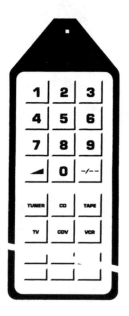

Figure 3.18 The shape of the remote control must indicate clearly how it should be held.

user and so on. In this two-way communication (dialogue) the following considerations apply:

- the user's expectations;
- the user-friendliness of the system;
- the help offered to the user when using the system.

Expectation

There must be a logical relationship between the method of operating a control and the user's expectation of the effect. An indicator in a car is, for example, always controlled from the same side of the steering wheel and its direction of use is always related to the intended direction of travel. If the position or direction is changed, it will cause great confusion.

Ensure compatibility in the direction of movement

The movements on a display must correspond to those of the controls (Figure 3.19). The following actions are the most natural in achieving a desired result:

On: upwards, to the right, away from the user, clockwise, pulled out.
Off: downwards, to the left, towards the user, anticlockwise, pressed in.
Increase: upwards, to the right, away from the user, clockwise, increasing resistance.
Decrease: downwards, to the left, towards the user, anticlockwise, decreasing resistance.

There are exceptions where the relationship is ambiguous. These are mostly due to ingrained habits. The direction of motion on a brake pedal in a car, for example, represents increasing break force, but also decreasing speed.

The objective of a control must be obvious from its location

The location of a control must bear a logical relationship to the position of the relevant information or to the effect. Softkeys, whose

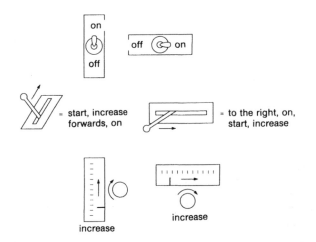

Figure 3.19 Movement of controls must correspond to display movements.

function is displayed on the screen directly above the keys, have already been discussed.

Another example is a stove, where the relationship between the location of the control knobs and that of the burners can be made to correspond, by placing the knobs on the top of the stove next to the burners in the same layout.

Use dual controls only when the consequences can be serious

Sometimes the number of controls, or keys, is smaller than the required number of control functions. In such cases additional action is carried out either before the control operation, after it, or during it. An additional action which precedes the control is called a prefix, one which follows the control is called a suffix. Known examples of a prefix are the keys * and # on a telephone or the CTRL and ALT keys on a keyboard. A suffix on such a keyboard would be, for example, the ENTER key. The use of prefix and suffix occurs at the expense of logic. The use of more than one prefix, or suffix, or the simultaneous operation of several controls is not recommended for normal circumstances, for example, do not use CTRL ALT F1. Prefix and suffix keys can be used to good effect if easy operation is undesirable; for example, in order to prevent accidental deletion of data. This is comparable to situations outside computing, where the use of both hands simultaneously is necessary to operate potentially dangerous equipment.

User-friendliness

Several principles are formulated to promote a user-centred approach to the development and evaluation of a product or system, so that the use of computerized interactive tools is enhanced by the improved quality of the dialogue. The aim is to increase the ease of use of the dialogue in terms of effectiveness, efficiency and satisfaction.

Make the dialogue suitable for the task

A dialogue is suitable for the task to the extent that it supports the user in the effective and efficient completion of the task. Some typical applications of this principle are:

• the system should present the user only with those concepts which are related to the user's activities in the context of the task;

- any activities required by the system but not related to the user's task should be carried out by the system itself;
- the type and format of input and output should be specified such that it suits the given task.

Make the dialogue self-descriptive

A dialogue is self-descriptive to the extent that each dialogue step is immediately comprehensible through feedback from the system, or is explained to the user on his requesting the relevant information. Some typical applications of this principle are:

- after any user action, the system should have the capability to initiate feedback;
- feedback or explanations should assist the user in gaining general understanding of the dialogue system;
- if defaults exist for a given task, they should be made available to the user, for instance, by giving hints (Figure 3.20).

Make the dialogue controllable

A dialogue is controllable to the extent that the user is able to maintain direction over the whole course of the interaction until the point at which the goal has been met. Some typical applications of this principle are:

- the speed of the operation should not be dictated by the system;
- if interactions are reversible and the task permits, it should be possible to undo the last dialogue step;

Graphics: Horizontal Line

1 – Horizontal position	Full
2 – Vertical position	Baseline
3 – Length of line	
4 – Width of line	0.013
5 – Grey shading (5% of black)	100%

Selection: 0

Figure 3.20 The system must provide hints to the user.

- the way that input/output data are represented should be under the control of the user, thus avoiding unnecessary input/output activities, such as entering 00123 where just 123 is more reasonable.

A dialogue should conform to the expectations of the user

A dialogue conforms to user expectations to the extent that it corresponds to the user's task knowledge, education, experience and to commonly accepted conventions. Some typical applications of this principle are:

- dialogue behaviour and appearance within a system should be consistent, the end of a command, for instance, should always occur in the same manner, for example always either with 'enter' or 'return' or simply nothing;
- the application should use the terms which are familiar to the user in the performance of the task; languages, e.g., English and Dutch, should not be mixed (Figure 3.21).

Make the dialogue error-tolerant

A dialogue is error-tolerant to the extent that, despite evident errors in input, the intended result may be achieved with either no or minimal corrective action having to be taken. Typical applications of this principle are:

- errors should be explained to help the user to correct them;
- the application should prevent the user from making errors;

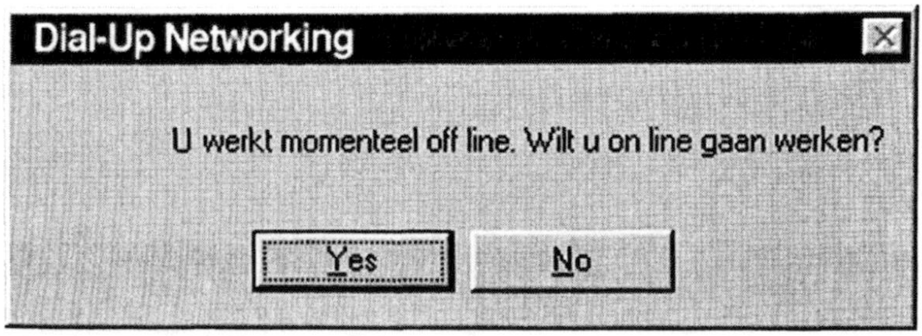

Figure 3.21 Using English programs within a Dutch operating system leads to bilingual dialogues.

• error messages should be formulated and presented in a comprehensible, objective and constructive style and in a consistent structure. Error messages should not contain any value judgements, such as 'this input is nonsense'.

Make the dialogue suitable for individualization

A dialogue is suitable for individualization to the extent that the dialogue system is constructed to allow for modification to the user's individual needs and skills for a given task. Some typical applications of this principle are:

• the amount of explanation (e.g., details in error messages, help information) should be modifiable according to the individual level of knowledge of the user;
• the user should be allowed to incorporate his or her own vocabulary to establish individual naming for objects and actions;
• the user should be able to modify the dialogue speed to his or her own working speed (for instance, the speed of scrolling information).

Make the dialogue suitable for learning

A dialogue is suitable for learning to the extent that it provides means, guidance and stimulation to the user during the learning phases. Some typical applications of this principle are:

• help information should be task-dependent;
• all kinds of strategies which help the user to become familiar with the dialogue elements should be applied; for instance, standard locations of messages and similar layout of screen elements. An example is the tip of the day in some programs (Figure 3.22).

 # Did you know this?

You can select more files or maps
by keeping the CTRL pressed, while
you click on the next files or maps

Figure 3.22 The 'tip of the day' helps the user to understand the program.

Different forms of dialogue

The way in which the exchange of information between people and computers is implemented in the software constitutes the dialogue form.

Not only computers are equipped with a VDU. Increasing numbers of other types of industrial systems include a small screen, sometimes with only a few lines (e.g., a mobile phone). Therefore, the rules described below for dialogue forms, with their advantages and disadvantages, are also applicable to such systems. The suitability of a given form depends on the application. A dialogue which is suitable for learning, such as described above, usually includes a combination of dialogue forms.

Use menus for users with limited knowledge or experience

With a menu, the user selects from a list of alternatives (Figure 3.23). The great advantage is that the user needs little knowledge or experience to understand the system. Furthermore, this type of dialogue requires little typing, it requires little mental effort and if a task is interrupted, the system immediately gives instructions on how to continue. The major disadvantage is that experienced users find it fairly long-winded and slow. In addition, the user has to read too much irrelevant information. Selections are made more rapidly from menus which have approximately seven topics.

If the menu does not fit on a page, a hierarchy, comparable to a family tree, has to be created. Because this means that only a small

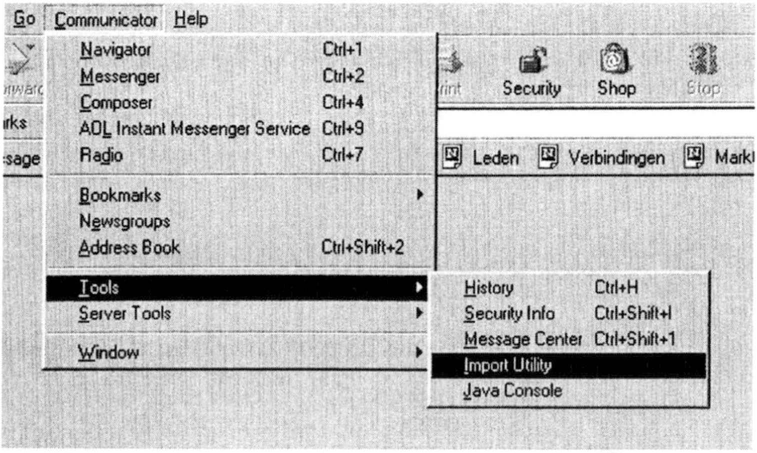

Figure 3.23 Menus do not require great mental effort on the part of the user.

portion of the hierarchy is ever shown on the screen, there is a chance that the user will lose his or her way. Complex hierarchies must be avoided.

Recognize the limitations of an input form

An input form consists of a number of filled-in screen sections (protected fields) which highlight the questions, together with some sections (free fields) to be filled in by the user to answer the questions. Here too, little training is required, as many users are familiar with paper form-filling. Just as with a menu structure, it requires barely any mental effort on the part of the user. The difficult manipulation of the cursor is a disadvantage because it may not be placed on the protected fields. What is more, the dialogue is not flexible: data must be filled in mostly in a fixed sequence.

Restrict the use of command language to experienced users

In command language, the user must key-in a fixed combination of characters to make the system react. The number of actions by the user is relatively small. To achieve the objective, the user therefore uses cryptic terms which are meaningless to a layman, for example CA.

The user must know the effect of the various input commands, and which commands are required to achieve a given effect, both of which require a considerable mental effort.

Direct manipulation must be consistent

A modern, frequently used form of dialogue is direct manipulation. This form is characterized by visual objects on the screen, where the user can point at, remove them, or change their value. A large part of the procedure with Apple MacIntosh or Microsoft Windows is based on this form of dialogue. Sometimes the method is inconsistent, like in the starting menu of Windows, where verbs and nouns are mixed.

Recognize the advantages and disadvantages of natural language

When controlling by natural language, the user communicates with the computer in the same manner as in normal conversation with people. This is the most familiar form of communication and gives the user a large degree of freedom.

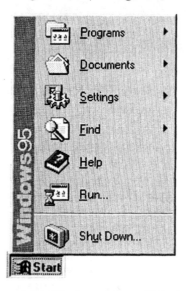

Figure 3.24 The Start menu of Windows is a mixture of unrelated terms.

Unfortunately, the disadvantages can often outweigh the advantages. Good understanding between people and computers requires that they speak the same language (for example Standard English). Natural language not only requires many more words than command language (Figure 3.25), but also, natural language is often vague and ambiguous. For example, in the sentence 'the man hit the girl with the umbrella' it is not obvious who was carrying the umbrella. Natural language can make the user overconfident in the system: 'if he understands this, then he must be able to do more'.

Computer: What train do you want?
User: I want to go to London this evening
Computer: Can you repeat the destination?
User: London
Computer: From Liverpool to London?
User: No
Computer: Where do you want to depart from?
User: Bristol
Computer: From Bristol to London?
User: Yes.

Figure 3.25 Speech dialogues require additional feedback.

Table 3.1 Help requirements depend on the user

Experienced user	Inexperienced user
help commands	help commands and keys
computer jargon	no jargon, or else, additional explanation
mathematical notation	examples
error messages	suggestions for corrections
no positive feedback	positive feedback
screen full of help	screen position also visible in addition to help
computer-oriented help	user-oriented help

Help

The user's need for help is usually considerable, because he or she did not design the (often complex) products and systems. Many manufacturers try to obviate this by supplying bulky instruction manuals along with their equipment. However, this does not appear to be the correct solution, as the manuals are seldom easily accessible and are often mislaid. Help should therefore be provided not only in paper form but also, wherever possible, on the system itself.

Match the type of help to the experience of the user

To be effective, the type of help must be suited to the experience of the user. Differences between the help requirements of experienced users such as programmers, and inexperienced users are shown in Table 3.1. Different types of help must be offered in parallel because systems are seldom accessed by only one type of user.

SUMMARY CHECKLIST

The user

1 Is the user-population defined as detailed as possible?
2 Are the mental models of the users taken into account?

Information

Visual information

3 Have texts with only capitals been avoided?
4 Has justifying text by means of blank spaces been avoided?
5 Have familiar typefaces been chosen?
6 Has confusion between characters been avoided?
7 Has the correct character size been chosen?
8 Are longer lines more widely spaced?
9 Is the contrast good?
10 Are the diagrams easily understood?
11 Have pictograms been properly used?
12 Has an appropriate method of displaying information been selected?

Hearing

13 Are sounds reserved for warning signals?
14 Has the correct pitch been chosen?
15 Is synthesized speech adjustable?

Other senses

16 Are taste, smell and temperature restricted to warning signals?
17 Is the sense of touch used for feedback from controls?
18 Are different senses used for simultaneous information?

Controls

Distinguishing between controls

19 Are differences between controls distinguishable by touch?
20 Is the location consistent and has sufficient spacing been provided?
21 Is unintentional operation avoided?
22 Are controls well within reach?
23 Are labels or symbols properly used?
24 Is the use of colour limited?

Types of control

25 Has the QWERTY layout been selected for the keyboard?
26 Has a logical layout been chosen for the numerical keypad?
27 Is the number of function keys limited?
28 Is the type of cursor control suited to the task?
29 Is the mouse used not too frequently?
30 Are touch screens used to facilitate operation by inexperienced users?
31 Are pedals only used where the use of the hands is inconvenient?
32 Are remote controls used to give the user more freedom?

Relationship between information and operation

Expectation

33 Is the direction of movement consistent with expectation?
34 Is the objective clear from the position of the controls?
35 Is dual control used only where the consequences can be serious?

User-friendliness

36 Is the dialogue suitable for the task?
37 Is the dialogue self-descriptive?
38 Is the dialogue controllable?
39 Does the dialogue conform to the expectations on the part of the user?
40 Is the dialogue error-tolerant?
41 Is the dialogue suitable for individualization?
42 Is the dialogue suitable for learning?

Different forms of dialogue

43 Have menus been used for users with little knowledge and experience?
44 Are the limitations of an input form known?
45 Has command language been restricted to experienced users?
46 Is direct manipulation consistent?
47 Have the disadvantages of natural language been recognized?

Help

48 Is the type of help suited to the level of the user?

4   Environmental factors

Physical and chemical environmental factors such as noise, vibration, lighting, climate and chemical substances can affect people's safety, health and comfort.

In this chapter we deal with these five factors in turn. Other environmental factors such as radiation and microbiological pollution (e.g., bacteria, moulds) are not discussed in this book. Guidelines on the maximum allowable exposure are given for each of these factors, followed by possible measures for reducing the exposure. In general three types of measure can be applied to reduce or eliminate the adverse effects of environmental factors:

- at source (eliminate or reduce source);
- in the transmission between source and Man (isolate source and/or Man);
- at the individual level (reduction of exposure duration, personal protective equipment).

Noise

The presence of high noise levels during a task can be annoying and, in time, result in impaired hearing. The first symptom of impaired hearing is a perceived difficulty in understanding speech in a noisy environment (party, pub, etc.). In such cases, a hearing aid is useless, as the background noises are also amplified. Annoyance, such as interference in communication or reduction of concentration, can occur even at relatively low noise levels. Annoyance and impaired hearing can be avoided by setting upper limits for noise levels. Noise levels are expressed in decibels, dB(A). Table 4.1 illustrates a few examples of noise levels.

Table 4.1 Examples of noise levels

Type of noise	dB(A)
jet engine at 25 m	130
jet aircraft starting at 50 m distance	120
pop group	110
pneumatic chisel	100
shouting, understandable only at short distances	90
normal conversation, understandable at 0.5 m	80
radio playing at full volume; loud conversation	70
group conversation	60
radio playing quietly; quiet conversation between two people	50
library reading room	40
slight domestic noises	30
gentle rustling of leaves; indoors in the evening in a quiet neighbourhood	20
very quiet environment	10
threshold of hearing	0

Guidelines on noise

The guidelines on noise given in this paragraph relate to the prevention of damage to hearing, as well as to the limitation of annoyance.

Keep the noise level below 80 decibel

A noise level which, over an 8-hour working day, exceeds 80 dB(A) on average, can damage hearing. Assuming constant noise levels, this daily level will be reached, for example, with an 8 hour exposure to 80 dB(A), or with a 1 hour exposure to 89 dB(A) (Figure 4.1).

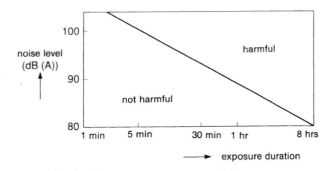

Figure 4.1 Constant noise levels should not last too long. To prevent impaired hearing, the average daily level should not exceed 80 dB(A).

Table 4.2 Maximum noise levels to avoid annoyance during various activities

Activity	dB(A)
Unskilled physical work (e.g., cleaning)	80
Skilled physical work (e.g., garage work)	75
Precision physical work (e.g., fitting and turning)	70
Routine administrative work (no full-time occupation)	70
Physical work with high precision requirements (e.g., fine grinding)	60
Simple administrative work with communication (e.g., activities in typing pools)	60
Administrative work with intellectual content (drawing and design work)	55
Concentrated intellectual work (e.g., working in office)	45
Concentrated intellectual work (e.g., reading in library)	35

Doubling the exposure time requires lowering the permissible noise level by 3 dB(A). If noise levels are variable, the average daily level is calculated from the individual noise levels. It is recommended that machines and suchlike be designed to keep the noise level at any time below 80 dB(A).

Limit the annoyance

Annoyance during thinking and communication tasks can already arise at levels well below 80 dB(A). An excess of noise will prove annoying even though the limit for damage to hearing has not yet been reached. It is mainly noise produced by others, unexpected noise and high frequency noise that cause annoyance.

Table 4.2 provides guidelines for the maximum allowable noise levels to avoid annoyance during various activities.

Rooms should not be too quiet

Although the aim is to reduce noise levels to below a certain maximum, at the same time, the level should not drop below 30 dB(A), otherwise unexpected irrelevant noise becomes too obvious.

Noise reduction at source

The most fundamental measures in noise reduction are ones taken at source. A number of possibilities for achieving this are discussed below.

Select a quiet working method

Consideration should also be given to noise levels when selecting a particular working method. A less noisy working method is not only of importance to those exposed to the noise; in many cases it also means less machine wear and less damage to the product. It is sometimes also possible to reduce or eliminate certain noisy stages in a process, for example finishing-off by grinding can sometimes be partly or totally eliminated.

Use quiet machines

Developments in the construction of 'quiet' equipment mean that an increasing number of quiet machines, tools and accessories have come onto the market. When selecting machines for purchase, attention should be paid to potential noise production during normal usage.

Well maintained machines are quieter

Poor fit, eccentricity and imbalance cause vibration, wear and noise. Regular maintenance of machines and equipment is therefore of great importance.

Enclose noisy machines

Noisy machines can be placed in a sound-insulating enclosure (Figure 4.2).

This can significantly reduce noise levels, but the disadvantage is that enclosed machines are less accessible for operation and maintenance. Special arrangements are also required to bring in and remove any process material, and possibly also to provide ventilation.

Noise reduction through workplace design, and work organization

Noise reduction is achieved in most cases by reducing or preventing the transmission of noise between source and receiver. A few relevant measures to improve the layout of the workplace and work organization are given in this section.

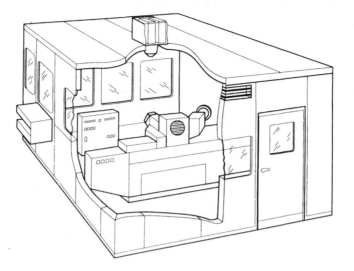

Figure 4.2 Noisy machines such as woodworking machines can be enclosed to reduce noise in the vicinity.

Separate noisy work from quiet work

Noisy activities can be segregated from quiet activities by having them carried out in separate areas, or outside normal working hours. The advantage is that fewer people are exposed to the noise, but other measures still have to be taken to protect those who are nevertheless exposed to the noise.

Maintain a sufficient distance from the noise source

The correct choice in selecting a location for a noise source is to keep it as far away as possible from those who may be exposed to it. Increasing the distance is most effective close to the source. For example, a given increase in distance, say of 5 m, from 5 to 10 m has more effect than an equal increase from 20 to 25 m.

Use the ceiling to absorb noise

The ceiling is often used to absorb noise. Although this decreases the noise level only to a limited extent, it is particularly effective in reducing the annoying effects, such as echo. Measures involving the ceiling are worth undertaking mainly in rooms where sound reverberates and where many workers are present. In existing buildings or in localized applications, one can use loose elements made of sound-damping material, which are hung from the ceiling. Another possibility is to install a lower ceiling made of such material. This also offers a way of concealing pipework, ducting, leads and suchlike, and can help thermal insulation.

Use acoustic screens

Acoustic screens placed between the source and the person can reduce the noise level. This measure is often only meaningful in combination with a sound-absorbing ceiling. The screen should be large enough to prevent the source of noise from being seen. Acoustic screens are ineffective if the distance between the source and the person is large.

Various types of screen are available: as a fixed wall, a moveable screen, a screen hung from the ceiling or one which can be attached to a machine.

Hearing conservation

One can resort to protecting hearing by using ear-plugs or -muffs if the previous measures, which were aimed at the source or at the transmission, are not feasible. Ear protectors must be available if the noise level is temporarily too high, for example during noisy maintenance activities. Different types of ear protectors are illustrated in Figure 4.3.

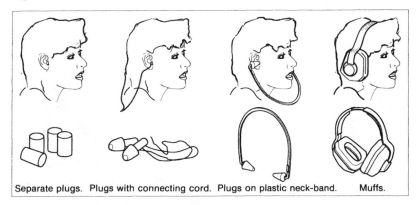

Separate plugs. Plugs with connecting cord. Plugs on plastic neck-band.   Muffs.

Figure 4.3 Examples of ear protectors. Ear protectors should only be used if other methods have failed to reduce noise sufficiently.

Ear-plugs are fitted into the ear, which also means that the degree of noise reduction is often limited if they are not properly used. Ear-muffs, by contrast, are placed over the ears. The resulting noise reduction is often greater than with plugs. They are also more convenient for frequent donning and removing, and they are more hygienic. Many people find them uncomfortable to wear because of perspiration, and they are also less suitable for those who wear spectacles. Noise reduction will be limited if the muffs do not fit closely over the ears.

Ear protectors must be suited to the noise and to the user

The pitch (frequency) of the noise must be taken into account when choosing ear protectors. Different types of protective equipment have maximum damping effects in certain frequency ranges. Data on the characteristics of ear protectors can be obtained from the suppliers. In order to encourage the use of ear protectors, personal preferences in comfort and ease of use must be taken into account. Different types of ear protectors should therefore be available.

Vibration

In any discussion of vibration, a distinction has to be made between whole-body vibration and hand-arm vibration. In whole-body vibration, the whole body is brought into vibration via the feet (in standing work) or via the seat (in seated work). Usually, the vibration is predominantly vertical, such as in vehicles. Hand-arm vibration affects only the hands and arms, and often arises when using motorized hand-held tools.

Three variables are important in assessing vibrations: their level (expressed in ms^{-2}), their frequency (expressed in Hz) and the exposure duration. Low-frequency body vibrations (<1 Hz), can produce a feeling of seasickness. Body vibrations between 1 and 100 Hz, especially between 4 and 8 Hz, can lead to chest pains, difficulties in breathing, back pain and impaired vision. The possible consequences of hand-arm vibration frequencies between 8 and 1000 Hz are reduced sensitivity and dexterity of the fingers, vibration 'white finger' (see below), as well as muscle, joint and bone disorders. The most common frequency range for hand-held motorized tools is between 25 and 150 Hz.

In practice, most vibration consists of several separate vibrations at different frequencies and in different directions. From the individual characteristics of these vibrations it is possible to calculate an average measure of the vibration level. This average level can then be used in practice to assess the impact of the vibration.

Guidelines on vibration

This paragraph contains guidelines on whole-body vibration, hand-arm vibration as well as shocks and jolts.

Body vibration should not result in discomfort

Body vibration results in discomfort at certain combinations of exposure duration and average vibration level. Figure 4.4 shows the limit for various combinations of exposure duration and average vibration level for whole-body vibrations in standing and seated work.

Prevent vibration 'white finger' resulting from hand-arm vibration

Vibration 'white finger' (also called 'dead finger') is caused by hand-arm vibration. The main symptom of this disorder is a reduction of blood flow in the fingers leading to discoloration of the skin. The fingers feel cold and become numb, which in time can actually lead to necrosis of the fingertips. The condition is aggravated by cold. The

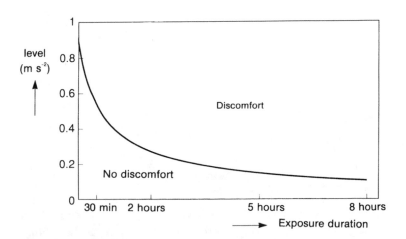

Figure 4.4 Body vibration results in discomfort, depending on exposure duration and average vibration level.

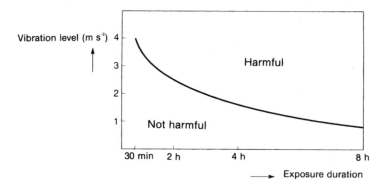

Figure 4.5 Hand-arm vibration can cause 'white finger', depending on the exposure duration and average hand-arm vibration level.

development of vibration 'white finger' depends among other things on the exposure duration and the average vibration level. Figure 4.5 shows the limit for various combinations of exposure duration and average vibration level.

Prevent shocks and jolts

Shocks and jolts often arise together with vibrations. Shocks and jolts with peak intensities more than three times higher than the average vibration level will increase the total vibration stress, and should be avoided. The reader should note that the guidelines for vibrations given here, assume that no shocks or jolts occur.

Preventing vibration

Vibration can be prevented at source, during the transmission between source and exposed person and, to a limited extent, at the individual level. In this section we deal with a few possible measures.

Tackle vibration at source

Large machines and motorized tools often constitute a source of vibration. Rotating movements generally cause less vibration than reciprocating movements, a fact worth remembering when designing or selecting machines and tools. Likewise, hydraulic and pneumatic transmissions are superior to mechanical transmission in this respect.

Heavy machines (those with a large mass) also generally cause less vibration.

Maintain machines regularly

Machines and hand-held tools sometimes display loose fits, eccentricity or imbalance, all of which cause vibration, noise and wear. Regular maintenance is therefore very important.

Prevent the transmission of vibration

Whenever measures at source are inadequate, attention should be devoted instead to reducing the transmission of vibration. This is best done by damping the vibration where it enters the body, for instance, by fitting floors, seats and handgrips with a damping material. An example is a well-damped seat in a bus, which makes it difficult for the vibrations to reach the body from the floor. The seating surface is fitted with a damping material and a pneumatic spring is located between seat and floor for damping.

If necessary, direct the measures at the individual

If measures at source and in transmission are not effective, then attention must be directed at the individual. This can be done by reducing the duration of exposure, for example by alternating tasks which entail vibration with tasks that do not entail vibration. Cold, humidity and smoking increase the risk of vibration 'white finger' and can be counteracted at the individual level, among other things, by using gloves for protection against cold and humidity.

Illumination

The light intensity, for instance, the amount of light which falls on the work surface, must be sufficiently high whenever visual tasks have to be carried out rapidly, and with precision and ease. Apart from light intensity, differences in luminance (contrast) in the visual field are also important. Luminance is the amount of light reflected back to the eyes from the surface of objects in the visual field.

Light intensity is expressed in lux, and luminance (brightness) in candela per m^2 ($cd\,m^{-2}$).

Guidelines on light intensity

In determining the amount of light which must fall from the surroundings onto a work surface, it is necessary to distinguish between orientation lighting, normal working lighting and special lighting.

Select a light intensity of 10–200 lux for orientation tasks

A light intensity of 10–200 lux is sufficient where the visual aspect is not critical, for example in the corridors of public buildings, or for general activities in store rooms, provided no reading is required. The minimum required intensity to detect obstacles is 10 lux. A higher light intensity may be necessary for reading noticeboards and the like, or to prevent excessive differences in brightness between adjoining areas; this allows the eyes to adjust more rapidly when moving between the areas, such as when driving into tunnels. Adjustment of the eye can take a fairly long time if the differences in brightness are large.

Select a light intensity of 200–800 lux for normal activities

Reading normal print, operating machines and carrying out assembly tasks can be considered normal visual tasks, and the following guidelines apply in this instance:

- a light intensity of 200 lux is adequate if the information is large enough and contrasts sufficiently with the background, for example, black letters printed on white paper (see Chapter 3);
- greater light intensities are necessary if the details are small or if the contrast is poor;
- people with limited vision and older persons require more light;

Table 4.3 Perception by humans of a few luminance ratios. Excessive differences in brightness within the visual field must be avoided

Luminance ratio	Perception
1	none
3	moderate
10	high
30	too high
100	far too high
300	extremely unpleasant

Luminance ratios greater than 10 are considered excessive.

- a greater light intensity is sometimes required to compensate for large differences in brightness between and within rooms, due, for example, to high light intensities in neighbouring rooms, or to the presence of windows.

Select a light intensity of 800–3000 lux for special applications

It is sometimes necessary to use localized task lighting. This can compensate for shadows or reflections on the work surface. For special activities such as visual inspection tasks, much higher illumination levels are used to enable fine details to be distinguished.

Guidelines on brightness differences

This section provides some guidelines on differences in brightness within the visual field.

Avoid excessive differences in brightness in the visual field

Excessive differences in brightness between objects or surfaces in the visual field are undesirable. Large differences can result among other things from reflections, dazzling lights and shadows. Table 4.3 shows a few examples of how people experience differences in brightness (expressed as the luminance ratio, which is the brightness of one object divided by the brightness of another).

Limit the brightness differences between the task area itself, the close surroundings, and the wider surroundings

The visual field can be divided into three zones: that of the task area, the close surroundings and the wider surroundings. The brightness of the task area should not be three times larger or three times smaller than that of the close surroundings. The brightness of the task area should not differ from that of the wider surroundings by more than a factor of ten. Differences in brightness that are too small should also be avoided because this makes a room look dull.

Improved lighting

Steps taken to improve lighting aim mainly to provide sufficient light intensity, and to avoid excessive brightness differences in the visual field such as may occur with light sources, windows, reflections and shadows.

Improve the legibility of information

When the visibility of the information is insufficient, it is more effective to improve the legibility of the information than to increase the light intensity. Further increases in light intensity are pointless when lighting is already intense. The legibility of information can be improved by enlarging the details (e.g., by using a larger typeface or smaller reading distance) or by increasing the contrast (e.g., black letters on a white background). Recommendations for the presentation of information are given in Chapter 3.

Select a combination of ambient and localized lighting

Except for orientation tasks, the required light intensity on a work surface can be achieved by a combination of fairly limited ambient lighting and more intense localized, or task lighting. The desired ratio between the general and the localized light intensity is determined among other things by the criteria on brightness difference between task and surroundings (see Guidelines on brightness differences, p. 85), and by personal preference. The intensity of any localized lighting must be adjustable.

Daylight can also be used for ambient lighting

Available daylight should also be used for general lighting. Incoming daylight and a view to the outside are much appreciated by most people. Large variations in daylight intensity from direct sunlight can be prevented by using blinds. Excessive brightness differences in the visual field (see p. 85) can occur in workplaces close to windows.

Screen sources of direct light

Blinding by direct light can be avoided by screening-off light which radiates sideways. However, vertical surfaces are then less well illuminated. This can be compensated for by opting for a light interior.

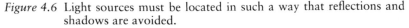

(a) incorrect   (b) correct

Figure 4.6 Light sources must be located in such a way that reflections and shadows are avoided.

Prevent reflections and shadows

Light sources must be located relative to the workplace so as to prevent reflections and shadows. Figure 4.6 shows the optimum location of light sources for a workplace. In VDU workstations, special care is required to prevent reflections on the screen.

Use diffuse lighting

Excessive reflections can be avoided by using indirect (i.e., diffuse) lighting in ceilings. Table tops, walls and suchlike should also produce diffuse reflection of any incident light, in other words, the reflected light should be evenly distributed in all directions. The surfaces should therefore have a matt finish. The optimum amount of light reflected from a surface (reflectance) depends on the purpose of that surface. Recommended reflectance values are given in Table 4.4. The reflectance is a value between zero and one, with a zero value meaning that no light is reflected (dark surface) and a value of one meaning that all the light is reflected (light surface).

Table 4.4 Recommended values for the reflectance of various surfaces

Surface	Reflectance
ceiling	0.80–0.90 ('light')
walls	0.40–0.60
table tops	0.25–0.45
floor	0.20–0.40 ('dark')

Flicker from fluorescent tubes can be avoided

Fluorescent tubes produce a flickering light which can be disturbing. It is possible to avoid any detectable flicker if two or more tubes are placed in a mounting so that they alternate out of phase; this is achievable by correct connection to the grid. The use of a single fluorescent tube, for example, to reduce power consumption, is ill-advised.

Climate

The indoor climate needs to satisfy several conditions if work is to be carried out in comfort. Four climatic factors (air temperature, radiation temperature of cold and hot surfaces, air velocity, and relative humidity) are significant in this respect. Whether a climate is considered pleasant depends also on the level of physical effort required by the work and on the type of clothing. Work is sometimes carried out in very cold environments such as in cold-rooms or outside, or in very warm environments such as near ovens. Special precautions are then necessary to prevent freezing or burning of the exposed skin, mainly on the face and hands. Without these precautions, the time spent in cold or hot environments has to be limited.

Guidelines on thermal comfort

This section contains guidelines on the four climatic factors of air temperature, air humidity, radiant temperature, and air velocity.

Allow people to control the climate themselves

Whether people find a climate pleasant depends very much on the individual. The aim must therefore be to allow people to control the climatic factors as far as possible themselves. This is feasible, for example, in an office with separate rooms.

Adjust air temperature to physical effort

Table 4.5 contains global guidelines on air temperature for tasks requiring different levels of physical effort. The guidelines ensure that people will feel comfortably cool to comfortably warm. The assumption here is that the air humidity is 30 per cent to 70 per cent, that the air velocity is less than $0.1\,\mathrm{m\,s^{-1}}$ and that normal clothing is worn.

Table 4.5 Guidelines on air temperature for tasks requiring different levels of physical effort

Type of work	Air temperature (°C)
Seated, thinking task	18–24
Seated, light manual work	16–22
Standing, light manual work	15–21
Standing, heavy manual work	14–20
Heavy work	13–19

Avoid very humid and very dry air

Humid air (relative humidity in excess of 70 per cent) or dry air (relative humidity less than 30 per cent) can affect thermal comfort. Dry air can lead to irritation of eyes and mucous membranes, and also increases the possibility of static electricity (risk of inflammation or ignition of chemical substances, unpleasant shocks, equipment failure). The humidity can be controlled either by adding moisture to the air or by removing it.

Avoid hot or cold radiating surfaces

Hot surfaces such as a roof and cold surfaces such as a cold window can affect thermal comfort. Steps must be taken whenever the radiant temperature of these surfaces differs by more than four degrees from the air temperature (see climate control, p. 90).

Prevent draughts

Draughts can affect thermal comfort, mainly in the case of light work. Draughts are uncomfortable at air velocities above $0.1\,\mathrm{m\,s^{-1}}$. Draughts can be caused among other things by ventilation (see pp. 91, 95).

Table 4.6 Materials which have to be touched should not be too hot if burns are to be avoided. The table shows the maximum temperature allowed depending on the exposure duration and the type of material

Duration of contact	Type of material	Maximum temperature (°C)
Up to 1 min	– metals	50
	– glass, ceramics, concrete	55
	– plastics (perspex, teflon), wood	60
Up to 10 min	– all materials	48
Up to 8 h	– all materials	43

Guidelines on heat and cold

Hot and cold environments are not only uncomfortable; hot climates, such as near ovens, can be energetically very stressful to the heart and lungs. In addition, parts of the body can be injured by burns or frost.

Avoid extremely hot and cold climates

Exposed parts of the skin can reach the threshold of pain in extremely hot climates or near very hot radiating surfaces. In a very cold climate, the hazard is frost-bite, the risk of which increases at high air speeds.

Materials which must be touched should be neither too cold nor too hot

If bare skin comes into contact with very cold metal, it may adhere to the metal surface. To be on the safe side, the temperature of metals likely to be touched should be at least 5 °C. A lower value can be tolerated for objects made of dry plastic or dry wood.

Table 4.6, on the other hand, shows the maximum temperature of materials allowed if skin burns due to contact are to be avoided.

Climate control

In this section we discuss measures relating to thermal comfort, and to hot and cold climates.

Locate equally heavy tasks together in a room

It is desirable that tasks which are more or less equally heavy should be located together in a separately heated room. This makes it possible to achieve a pleasant climate for each group of tasks.

Adjust outdoor tasks to the climate

It is not possible to control the outdoor climate, but cold and hot outdoor climates, to a certain extent, can be tolerated better by adjusting the energy demand of the task. In a cold climate, tasks should be heavier so as to increase body temperature and reduce the risk of freezing. In a hot climate, the opposite applies.

Adjust the air velocity

Where there is a draught (maximum $0.1\,\mathrm{m\,s^{-1}}$) it is sensible to increase the air temperature to allow work to be carried out in comfort. In very cold climates the air velocity should always be as low as possible to prevent parts of the body from freezing. Conversely, a hot environment becomes more pleasant if the air velocity increases.

Prevent unwanted hot or cold radiation

Hot or cold radiation can be suppressed by insulating or screening off hot or cold radiating surfaces such as walls, floors, roofs and windows. In addition, correct layout of the workspace can help increase the distance between the person and the source of radiation. Finally, the air temperature can also be adjusted to reduce the difference between air temperature and radiant temperature.

Limit the time spent in hot or cold environments

People themselves should be able to determine how much time they spend in hot or cold environments.

Use special clothing when working for long periods in hot or cold environments

Clothing with a high insulation value affords protection against cold. Similarly, special clothing can also be used as protection against heat (e.g., firemen).

Chemical substances

Chemical substances occur in the environment as liquids, gases, vapours, dusts or solids. Some substances can cause discomfort or present a health hazard if inhaled, ingested or if they come into contact with the skin or eyes. The symptoms can develop immediately or at a later stage. It is known that many substances are irritants, carcinogens, mutagens (damage genes) or teratogens (lead to birth defects). The body must therefore be exposed as little as possible to such chemical substances.

Guidelines on chemical substances

The most important guidelines on chemical substances, given in this paragraph, are based on the so-called TLVs (threshold limit values). These are official international limits for chemical substances in air, and are intended to prevent adverse health effects (rather than discomfort).

Apply TLVs or other limits as maxima for chemical substances in ambient air

TLVs are available for several hundred substances. The list of TLVs is regularly updated by the inclusion of new substances and by taking account of new data on the toxicity of substances.

The TLV is an 8-hour weighted average concentration, and should not be exceeded in any single day.

Certain substances have a rapid toxic effect, in which case a separate TLV is established, namely the TLV-C (C = ceiling) which may not be exceeded at any time (see below). Only a small proportion of the known chemical substances appear in the TLV list. Whenever a chemical substance does not appear in a national TLV list, the lists of other countries or toxicological handbooks can be consulted. If the substance appears in none of these, it still does not mean that it is harmless. In this event, individual organizations often tend to apply their own standards.

Avoid carcinogenic substances

Certain airborne substances are known to cause cancer. Exposure to these substances must be avoided at all times. Table 4.7 shows a random selection of chemical substances which could be present in air and are considered to induce cancer. A comprehensive overview of known or suspected carcinogens can be found in publications of

Table 4.7 Random selection of chemical substances which are considered to induce cancer

Substance	Example of use
Asbestos	Thermal insulation
Benzene (benzol)	Solvent
Chrome compounds	Pigment
Polycyclic hydrocarbons	Component of tar
Vinylchloride	Raw material for PVC

the International Agency for Research on Cancer (IARC), Lyons, France.

Avoid peak exposures

Short-term exposure to high concentrations of a chemical substance can affect health even if the TLV is, on average, not exceeded over an 8-hour period. Therefore, the TLV-C values should be applied for substances with a rapid toxic effect. The design of a work or living environment must ensure that the TLV-C is not exceeded in other circumstances, such as cleaning or maintenance.

Exposure to mixtures of substances should be limited

Although in practice one is confronted mainly with mixtures of substances, there are generally no TLVs for such situations. There is also no guarantee that satisfying the individual TLVs will avoid health risks, as the effects of the individual substances can reinforce each other.

Always aim to remain as far below the TLVs as possible

It is important to try to remain as far as possible below the TLVs at all times. The rule of thumb in the design of new work or living environments is to achieve concentrations of less than a fifth of the TLVs. Remember also that remaining below the TLVs does not guarantee the absence of any discomfort (e.g., pungent smells). Conversely, substances which cause no discomfort can in fact be dangerous.

Packages of chemicals should be labelled appropriately

The supplier of a chemical substance must provide information on the toxicity of the substance and how to use it. The first indication of this must appear on the label, which should bear appropriate standard warning signs (Figure 4.7).

Measures taken at source

Measures can be aimed at the source (this section) or at the exposure (see following section). Measures at source are preferable, especially if they mean replacing the source. If this is not feasible, the source should be reduced. If this is still inadequate then the source must be

explosive

flammable

corrosive

oxidizing

toxic

harmful or irritating

Figure 4.7 Warning signs for chemical substances.

isolated. Measures taken at source can be directed at the chemical substance itself, the (production) process, or the working method.

Remove the source

A fundamental measure at source is to replace the harmful substance by substances which, so far as is known, are not harmful, or at least less so. Examples of this are the use of water-soluble paints instead of solvent-based paints, and thermal insulation using rock-wool rather than asbestos. Harmful production processes or working methods must be replaced by processes or working methods which are less harmful. An example of this is the use of an industrial vacuum cleaner instead of compressed air in cleaning activities.

Reduce the emission at source

The reduction of emission at source can affect the chemical substance itself, the production process or the working method. Examples of measures aimed at the substance itself are the use of paint with a lower concentration of heavy metals, and the supply of raw materials in the form of a paste rather than a powder. Examples of measures aimed at the production process are: reducing emissions by carefully tuning the process, carrying out regular maintenance, and reducing the fall height when emptying sacks of powder. An example of a measure aimed at the working method is to allow painted workpieces to dry in a separate room instead of in the spray cabinet.

Isolate the source of chemicals

A third measure aimed at the source is to prevent harmful substances from being released. An example is the use of enclosed, instead of open, transport systems for material, thus preventing the release of harmful chemicals.

Ventilation

Measures aimed at the exposure route should be taken whenever those directed at the source are inadequate. In this section we discuss measures directed at the transfer between the source and people (ventilation of the air). The section that follows addresses measures aimed at individual exposure, such as organizational measures, or use of personal protective equipment.

Chemical substances must be extracted directly at source

If it is not possible to prevent the release of chemical substances, then harmful substances should be extracted directly at source. In such instances, the extracted air is often released to the environment without being cleaned. However, environmental laws restrict the permissible concentration of chemical substances in the exhaust air, and as these laws become more stringent, this will stimulate measures aimed at the source. Note that if harmful substances in air can be partly removed through an exhaust system, this at the same time requires fresh air to be supplied to the workplace. Figure 4.8 shows an example of an exhaust system used in spray cabinets.

Provide an efficient exhaust system

In practice the effect of exhaust systems is often disappointing because of less than ideal conditions, such as unfavourable location, inadequate maintenance, or faults. Avoid extracting the air only from higher up in the room, rather than from the breathing zone. Regular maintenance is essential to prevent dirt from reducing the efficiency of the system.

Figure 4.8 Use of an exhaust system in spraying activities, which removes residual paints and solvents from the air inhaled by the worker.

Pay attention to the effect on climate when designing air extraction and ventilation

Air extraction and ventilation increase the chance of a draught. This in turn influences the degree of thermal comfort (see page 88). It is also important that the fresh air supply be pre-heated.

Provide sufficient air changes

Indoor environments must also be adequately ventilated even if no dangerous substances are present. The required volume of fresh air per person and the rate of air change depend on the degree to which the work is physically demanding (Table 4.8).

Table 4.8 Recommended optimum space and air change

Nature of the work	Volume per person (m³)	Fresh air supply rates (m³ hr⁻¹)
Very light	10	30
Light	12	35
Moderate	15	50
Heavy	18	60

Measures at the individual level

Measures to reduce the effect of chemical substances at the individual level are either organizational, whereby persons are exposed for as little time as possible to the substances, or else involve the use of personal protective equipment.

Implement organizational measures

Various organizational measures are possible to reduce people's exposure. People should spend as little time as possible in rooms with contaminated air, and likewise, the number of people exposed to the contaminated air should be limited. Activities where chemical substances are released can, for example, be separated from unaffected activities by locating them in a separate room. It is also feasible to carry out these activities outside normal working hours. The advantage is that fewer people are exposed to the substances. However, other precautions must be taken to protect those who are nevertheless exposed.

Use personal protective equipment

If measures aimed at the source or the exposure are not feasible or adequate, then personal protective equipment must be used, even though most users consider such equipment to be a nuisance.

In emergencies it is possible to use special masks fitted with filters, as these can provide protection against a number of gases. Special masks can also be used against fine dust. The masks must closely fit the shape of the face and make proper contact with the skin. Instruction in the use of masks is essential.

Use dust masks only for protection against coarse dust

Dust masks offer no protection against gases. They can retain part of the coarse, chemically harmless dust particles, but are inadequate at high concentrations (for example, if a mist hangs in the room).

Use protective equipment and gloves

Adequate protective equipment, such as gloves and aprons should be worn when working with liquids which can be absorbed through the skin. Gloves are often considered to be a nuisance, but on the other hand, the protective effect of special skin creams is unproven.

Ensure a high standard of personal hygiene

Other kinds of measure can be taken to reduce the absorption of chemical substances through the skin:

- clean dirty clothing and gloves regularly;
- do not use dirty cleaning cloths;
- cleanse the skin regularly with soap and water;
- ensure rapid treatment of skin lesions.

SUMMARY CHECKLIST

Noise

Guidelines on noise

1 Is the noise level below 80 decibel?
2 Is annoyance due to noise avoided?
3 Are rooms perhaps too quiet?

Noise reduction at source

4 Has a low-noise working method been chosen?
5 Are quiet machines used?
6 Are machines well-maintained?
7 Are noisy machines enclosed?

Noise reduction through workplace design and work organization

8 Is noisy work separated from quiet work?
9 Is there an adequate distance to the source of noise?
10 Is the ceiling used for noise absorption?
11 Are acoustic screens used?

Hearing conservation

12 Are hearing conservation measures suited both to the noise and to the user?

Vibration

Guidelines on vibration

13 Is uncomfortable body vibration avoided?
14 Is vibration 'white finger' due to hand-arm vibration avoided?
15 Are shocks and jolts prevented?

Preventing vibration

16 Is vibration tackled at source?
17 Are machines regularly maintained?
18 Is the transmission of vibration prevented?
19 Are measures at the individual level applied only as a last resort?

Illumination

Guidelines on light intensity

20 Is the light intensity for orientation tasks in the range 10–200 lux?
21 Is the light intensity for normal activities in the range 200–800 lux?
22 Is the light intensity for special applications in the range 800–3000 lux?

Guidelines on brightness differences

23 Are large brightness differences in the visual field avoided?
24 Are the brightness differences between task area, close surroundings and wider surroundings limited?

Improved lighting

25 Is the information easily legible?
26 Is ambient lighting combined with localized lighting?
27 Is daylight also used for ambient lighting?
28 Are light sources properly screened?
29 Are reflections and shadows avoided?
30 Is diffuse light used?
31 Is flicker from fluorescent tubes avoided?

Climate

Guidelines on thermal comfort

32 Are people able to control the climate themselves?
33 Is the air temperature suited to the physical demands of the task?
34 Is the air prevented from becoming either too dry or too humid?
35 Are hot or cold radiating surfaces avoided?
36 Are draughts prevented?

Guidelines on heat and cold

37 Is the climate possibly too hot or too cold?
38 Are materials which have to be touched neither too cold nor too hot?

Climate control

39 Are equally heavy tasks grouped in the same room?

40 Are the physical demands of the task adjusted to the external climate?
41 Is the air velocity optimal?
42 Are undesirable hot and cold radiation prevented?
43 Is the time spent in hot or cold environments limited?
44 Is special clothing used when spending long periods in hot or cold environments?

Chemical substances

Guidelines on chemical substances

45 Is the concentration of chemical substances in air subject to limits (TLVs or other)?
46 Are carcinogenic substances avoided?
47 Are peak loadings prevented?
48 Is exposure to mixtures of substances prevented?
49 Is the concentration of chemical substances as far below the TLVs as possible?
50 Does the labelling on packages of chemicals provide information on the nature of any hazard due to the contents?

Measures taken at source

51 Can the source be removed?
52 Can releases from the source be reduced?
53 Can the source be isolated?

Ventilation

54 Are chemical substances extracted directly at the source?
55 Is the air exhaust system efficient?
56 Has attention been paid to climate at workplaces where exhaust and ventilation are used?
57 Are sufficient air changes provided?

Measures at the individual level

58 Are organizational measures possible?
59 Is personal protective equipment available?
60 Are dust masks used only as protection against coarse dust?
61 Are protective clothing and gloves available if necessary?
62 Is attention paid to personal hygiene?

5 Work organization, jobs and tasks

Activities of human beings usually take place in a wider organizational context. The activities of one person are related to the activities of others. Usually an organization is divided into units. An example is a post office. This office has different organizational units, for instance concerning sales, mail delivery and financial matters. Within a unit people are employed in certain jobs. Let's consider a job at the stamp window. The employee has different tasks like selling stamps and paying cheques. To perform the task to sell stamps, different actions are needed, like handling the stamps, cashing the money and returning the change. The relation between jobs, tasks and actions is presented in Figure 5.1.

In this chapter recommendations are given for the design of work organization, jobs and tasks. The focus is primarily on paid work. However, most of the guidelines are also applicable outside of work (household, sport, voluntary work, traffic).

Tasks

The design of work organizations, jobs and tasks always begins with a description of all the tasks to be performed.

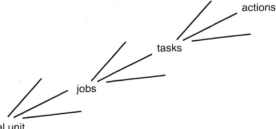

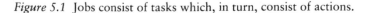

Figure 5.1 Jobs consist of tasks which, in turn, consist of actions.

Describe tasks in a neutral way and not the way they are performed

There are different ways to perform a task. The task 'selling stamps' can be done by a human being as well as by a machine. Both humans and machines can perform the tasks in different ways (for instance, the procedure to cash money or the type of machine involved).

When dividing between humans and machines, keep in mind their specific qualities

After the tasks have been determined they have to be distributed between humans and machines. This process is called allocation of tasks. Some tasks can be better performed by humans, others by machines. The following guidelines can be used in deciding when to allocate tasks to people rather than machines.

- People are more creative than machines in solving problems, especially unexpected ones.
- People can communicate by several means, such as speech, visual expressions or gestures.
- People would rather not allocate certain tasks to machines, for instance, weighing alternatives in order to make decisions.
- People are sometimes cheaper than machines. This is particularly true for complex movement patterns which are required only occasionally.
- People are better at filtering relevant information from a mass.
- Machines are better at counting and repetitive actions.
- Machines can operate in unhealthy and extreme conditions.

Economic factors often play a crucial role in deciding the allocation of a task which is otherwise equally suited to people or machines.

Jobs

After the allocation of tasks, the tasks have to be combined into interesting jobs for people. The following criteria must be met if a job is to be interesting:

- completeness of the job;
- control over the work;
- absence of repetitive tasks;

- alternation of difficult and simple tasks;
- freedom of workers to determine the method, the sequence of operations and pace within the job;
- opportunity to make contact with others;
- availability of information.

These attributes are discussed in detail below.

Jobs must consist of more than one task

A job is complete if it consists of a logical, coherent group of preparatory, production and support tasks.

Opportunities to learn and control will arise mainly through a job's support and preparatory tasks. The terms 'job enrichment' or 'vertical job enlargement' are used to describe the process of making a job more complete. By contrast, assigning more of the same type of tasks to a job is called 'job extension' or 'horizontal job enlargement'.

Everyone contributes to solving problems

Helping to solve problems makes a job interesting. This is particularly true if the problems go beyond run-of-the-mill difficulties, to include the unexpected. Solutions are usually arrived at through consultation, which can be of the following nature:

- functional consultation: maintaining contacts with immediate colleagues;
- work consultation: regular consultation with colleagues and line management;
- working parties: the solution of particular problems by separately formed groups.

A prerequisite for all these forms of consultation is that not only should discussion take place, but agreement on solutions must be reached; furthermore the problems should be such that solutions do exist.

The cycle time must exceed one and a half minutes

Many jobs require repeating the same type of actions. This is most clearly illustrated by conveyor belt work. The time between two repeats, the cycle time, should not be too short, preferably not less than one and a half minutes. Shorter tasks are totally mind-numbing and should not constitute the main component of a job. The ability to influence the task is largely absent in short cycle tasks: the work pace, the location, as well as the starting and finishing times, are all fixed.

Alternate easy and difficult tasks

A job should be made up of both simple and difficult tasks. If a job consists entirely of difficult tasks, there is a risk of mental overstress. If there are too many easy tasks, the worker will not feel challenged and boredom will set in.

Allow people to decide independently how to do their work

People will find a job more interesting if they can decide independently how to do their work. This autonomy can relate to the method of working, the order in which the individual actions are carried out, and the place of work. Other aspects are the possibility of rejecting supplied material (e.g., substandard raw materials, components, information) and the authority to call on the help of others.

Provide contacts

Contacts with others must form a component of the job. A contact can take place in several ways:

- by helping one another;
- by discussing the work;
- by talking of things other than work.

The job can be made more interesting by widening the diversity of contacts or extending the time spent making these contacts.

Informal contacts take place when the employee needs to leave his or her workplace once in a while or when employees work in close proximity. A noisy environment adversely affects opportunities for contact.

Tasks should be accompanied by sufficient information

A sustained flow of information is needed if the best use is to be made of the opportunities people have of properly controlling their tasks. This implies information at two levels, namely the workplace level, and the division or company level.

The type of information can also be of a dual nature:

- feedback: people subsequently receive information on the quality and quantity of work produced;
- forward coupling: prior information on the required quality and quantity, as well as on the influence of the pace of work on these two factors.

This information must help highlight the objective and the results of the work.

Work organization

When combining jobs in a work organization the following aspects are relevant:

- flexible forms of organizations;
- autonomous groups;
- coaching management styles.

These aspects will be described below. A special focus is put on the design of autonomous groups and the new role of the management to support the optimal performance of the employees within the new developments.

Flexible forms of organizations

Organizations must be able to react quickly to changing environments and an ongoing process of renewal of products, services and labour processes.

Replace hierarchical work organizations by more flexible structures.

The traditional hierarchical work organization is increasingly replaced by more flexible structures (Figure 5.2).

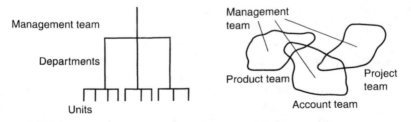

Figure 5.2 Hierarchical structures are replaced by more flexible forms.

The layers of the organization disappear (more flat organization) and the boundaries between the organizational units become blurred (more co-operation to serve the customers). More tasks and responsibilities are allocated to employees lower in the organization. Workers work together in self-supporting teams to use each other's strengths. The role of the manager changes from 'boss' to 'coach'.

To change an organization in this way, bravery and persistence are needed from the management. It is attended by a change of culture: another way of communication, control and justification. Extra attention is needed for education, training and implementation. External support can be needed to guide these processes.

Make housing conditions and working times flexible

A flexible form of organization means that employees are judged less on their presence at work, but more on the result of their activities. This means that workers should not always have their own individual office (or room) but that the space is used more flexibly. Sometimes it is possible to perform part of the work activity at home, instead of at the workplace. In an office environment this could mean that workers take the first free space when they arrive. Their personal belongings can be taken in a movable drawer. This flexibility also means that working and resting hours can be taken more flexibly and that they can be determined by the social environment or by the traffic conditions. Of course, this form of organization is not applicable to everyone; in industry the occupation of machines and the dependency of suppliers plays an important role.

Autonomous groups

Teamwork is an alternative for the fixed forms of organization with individual control of employees. An autonomous group is a fixed group of employees who, together, are responsible for the total process in which products or services are realized, without continuously consulting a manager.

This form of organization can contribute to shorter production times, higher productivity, higher quality, more innovations, higher flexibility, improvement of the quality of work and better labour relations. Above all production risks are smaller, because people are in more than one position in the team. Some guidelines for the operation of groups follow.

The assignment to the group must be clear

The assignment to the group must consist of strongly related activities. The assignment must be clear and the result must be identifiable and measurable. The starting situation is described precisely and the members of the team depend on each other. The team has sufficient possibilities to make decisions: decide within the conditions how the product is realized. An example is the co-operative work on the interior of a car (see Figure 5.3).

The team size must be 7–12 members

The optimal size of teams is 7–12 members. The size is determined by:

- the involvement of team members;
- the time span in which decisions can be taken;
- the productivity;
- the ability of the group to solve problems.

In bigger teams, the productivity and ability to solve problems will be better and the other aspects worse. Smaller teams need less time to decide and have more involvement.

Coaching management styles

In a flexible organization where the responsibilities are lower in the organization, the role of the management changes.

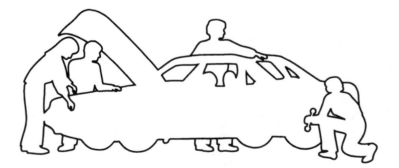

Figure 5.3 The group must work towards a recognizable goal.

Table 5.1 The difference between boss and coach

The Boss
- The boss does not tell everything, he keeps some information back;
- the boss knows everything, you can ask him everything;
- the boss knows everything better, and clearly shows this to be so;
- the boss solves the problems;
- the boss is always busy with his own work;
- the communication between boss and employee is restricted to work instructions;
- if you work well you hear nothing, when it goes wrong you are in trouble;
- when things go wrong it is always a mistake of the employee.

The Coach
- The coach presents information to those involved as soon as possible;
- the coach does not need to know everything;
- the coach and the employee depend on each other;
- the coach facilitates employees to solve their own problems;
- the coach supports the work of the employees;
- the communication between coach and employee is two-way;
- the coach is interested in the employee; he asks questions and listens;
- when things go wrong the coach asks himself what to do to prevent mistakes.

Don't act as the boss, but coach the employees

In the new management style that fits flexible organizations better, the manager is no longer the 'boss' but acts as 'coach'. The differences between these styles of management are presented in Table 5.1.

SUMMARY CHECKLIST

Tasks

1 Are tasks described in a neutral way?
2 Has a conscious decision been made about allocating tasks to a person or to a machine?

Jobs

3 Does the job consist of more than one task?
4 Do those involved contribute to problem solving?
5 Is the cycle time longer than one and a half minutes?
6 Is there alternation between easy and difficult tasks?
7 Can those involved decide independently on how the tasks are carried out?
8 Are there adequate possibilities for contact with others?
9 Is the information provided sufficient to control the task?

Work organization

Flexible forms of organization

10 Are hierarchical work organizations replaced by more flexible structures?
11 Are housing-conditions and working times flexible?

Autonomous groups

12 Is the assignment to the group clear?
13 Does the group consist of 7–12 members?

Coaching management styles

14 Is the role of the manager more coach than boss?

6 The ergonomic approach

An ergonomic approach can be adopted in virtually any kind of design or purchasing project. Such an approach merely requires the systematic application of ergonomic principles. The person providing the ergonomic input to the project, whom we will refer to as the 'ergonomist', must work systematically, and wherever necessary should call upon other specialists. In this chapter we present an appropriate general methodology for such an approach, which might be applicable to the following types of project:

- selecting a commercially available product for purchase;
- improving an existing product or system;
- designing a new product or system;
- adapting an individual workplace;
- refurbishing a business or workplace, for instance, after automation;
- designing a complete plant.

These guidelines do not provide a universal methodology because of the wide variety of possible projects. Thus for many projects some of the guidelines will be omitted because they are not applicable. The recommendations given here are addressed to the ergonomist, but when he is part of a team, they apply also to the other team members. The specific roles which the ergonomist can fulfil are those of:

- subject specialist;
- intermediary between technical designers, users and management.

The chapter concludes with a control list, or so-called checklist, which can be used to ascertain that no ergonomic aspects of importance to the project have been neglected. The checklist is based on the knowledge gained in the previous chapters.

Project management

Involve users in the project

An important characteristic of the ergonomic approach is the involvement of users and other stakeholders in a project, in the earliest possible phase ('participatory ergonomics'). The goal of this policy is:

- avoidance of mistakes in design or purchase;
- enlargement of acceptation;
- development of more ideas;
- faster identification of bottlenecks;
- give workers a say in their work and enlarge both autonomy and human well-being.

This procedure is especially important when many people are going to use the design or product, if a wrong design has serious consequences or when there is mistrust and rejection.

The method is less applicable when the project has to be finished fast, when the management is not open to the principle of participation or when there are no significant advantages (for instance a solution without any human workers).

For a participatory ergonomics approach, guidance of an ergonomist is desirable to bring in ergonomics knowledge, to promote stakeholders' co-operation and to achieve concrete results.

Introduce ergonomic requirements as early as possible in the project

It is better to apply ergonomics from the outset (prevention) rather than retrospectively (cure). Ergonomic requirements should therefore be introduced as early as possible in a project, and must play a role in every one of its phases. It is still all too often the case that ergonomists are only involved when a project or system is almost completed. They are then expected to append a sprinkling of ergonomics, which means that a fundamental contribution is out of the question.

Use conventional methods for project management

In order to contribute effectively to a project, the manner in which the ergonomist provides an input must be suited to the method which is usual for the project team. Design or purchasing usually involve the following separate, successive stages (Figure 6.1).

a   Initiative: making a detailed survey of the project, formulating the questions and planning the rest of the project.
b   Problem identification: gathering the required data, which usually leads to a multitude of alternative solutions.
c   Selection of solutions: selecting from the alternatives and further developing the selected option.
d   Implementation: implementing the selected option.
e   Evaluation: evaluating the option and the project.

These stages are described in more detail in the next five sections (pp. 114–122).

Make sure that the planning is flexible

The role of planning in any project is only to provide a logical sequence and to avoid aspects being overlooked. In practice there will always be an iterative process where any phase can lead to a revision of the previous one. For example, if in gathering data it appears that the project is still not properly defined, then a reformulation of the project becomes necessary.

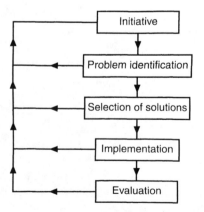

Figure 6.1 The successive stages of a design or purchasing project.

114 *Ergonomics for Beginners*

Realize that co-operation implies joint responsibility

All those who work on a project share the responsibility for the final result. During a project an ergonomist must regularly ask himself whether it is worthwhile to continue contributing. It can happen that the ergonomist has to compromise to such an extent that the label 'ergonomically designed' is not really applicable to the end-product.

Initiative phase

A project starts with a given objective, but this is seldom formulated in such a way that ergonomists can be involved immediately. Therefore we must attempt to describe the objective clearly and in a detailed manner. This requires a series of discussions with all parties.

Define who is involved in the project

It is important to the success of a project to know exactly who is involved in a project and who will have to deal with the result. The parties are:

- clients (e.g., company managers);
- members of the project team;
- help brought into the project, for example, subcontractors and experts;
- users and their environment.

As soon as the names are known, it must be made clear how consultation will take place and how binding certain opinions or advice will be. The future users are best involved through a steering committee. The information on the project must be understandable to every person involved.

Make sure that those involved support the project

The support of all those involved in a project is essential to its success. In a business, this includes managers and workers, as well as suppliers. Their support also has a bearing on the course of the project and its outcome. Enthusiasm on their part should be cultivated by a good presentation at the beginning of the project, describing its course and potential benefits.

Agree a code of conduct

It must be clear at the outset what freedom the ergonomist and other team members have in the project, such as access to rooms and documents. Agreement must also be reached on how the users consulted by the ergonomist or involved in a survey, are informed of the result. This affects, for example:

• access to documents and/or parts of the company;
• possibilities and limitations for publicizing the future results.

Ergonomists must always treat information about individuals in confidence. Indeed, even the client is given no access to the survey data on individuals.

Do not raise false expectations on the part of the users

Whenever an ergonomist approaches users in a project, the contact with them must be by prior arrangement. It is also worth bearing in mind that people who are dissatisfied, sometimes expect the visit of an ergonomist to lead to immediate improvements. What is more, users sometimes associate such a visit with purely coincidental reorganization measures, such as dismissals or transfers. Care is certainly of the greatest importance when the aim of a visit is to obtain comparative data for a project, from a similar part of the company.

State the limits of the project

It is mainly in production systems that a change in a section (e.g., a division or department) will have an influence on other sections. The boundaries of the project must therefore be properly established; these depend on available finance and time. In addition, management plans for the future must be made known.

Describe the planned course of the project at the outset

The intended course of the project must be absolutely clear from the start. Agreements about the evaluation phase must also be made at the beginning: when, how and with whom the evaluation will take place. Possible shortcomings in the solution can be identified through this evaluation, which usually also provides a retrospective view of the adopted working method. The latter can be useful in future projects of a similar nature.

Problem identification phase

The problem identification phase is aimed at gathering the relevant ergonomic aspects of the project on paper. It involves compiling complaints, ideas and wishes, and subsequently analysing the ergonomic aspects in the light of:

• safety, health and comfort associated with the adopted solution;
• performance, in other words, the usability of the adopted solution.

The usability is determined both by objective and subjective elements:

• objective: the efficiency (can the user work rapidly and without error with the product or system?);
• subjective: the acceptance (does the user wish to work with the product or system and does its use not lead to stress?).

Use a checklist to avoid overlooking any aspects (see p. 123), and ensure that the aspects are not only examined and assessed in isolation, but also in the context of the whole project.

Establish at the start how the data will be processed

Prior to gathering data it is necessary to reflect on the manner of its subsequent processing. The method of data processing must be tested beforehand if large amounts of data need to be compiled. It is also sensible at this point to prepare for a possible evaluation survey, which may involve, for example, the same persons having to be approached again.

Select more than one analysis technique

Various techniques can be used to gain an insight into the ergonomic aspects of the project; selecting the technique depends on the specific circumstances. It is generally sensible to use different techniques side by side in order to obtain a complete and reliable picture. Here are a few techniques:

• analysis of documents and statistics, such as user statistics, absenteeism data, and registered complaints. These provide a first impression of the project;

- observations: relevant events such as tasks and operations are observed;
- interviews: an impression is gained of the users' experience with problems through a questionnaire which is more or less structured, depending on circumstances;
- group discussions: the problem is discussed with a limited group of users (say 6–12 people);
- written questionnaire: data are gathered from large groups of users;
- experimental methods: some ergonomic aspects are investigated in a controlled manner in a laboratory or in the field. The data are obtained by measurements on people or on their physical environment.

The specific choice and effect of these techniques depends on the type of project.

Always start with a survey of existing documents

There is no point in investigating aspects which are sufficiently well described in existing documents. Therefore the ergonomic contribution should always start with a survey of all existing relevant information, which will include both project documentation and specialist ergonomic literature. Good overviews of the more recent general ergonomics literature are available (see also Chapter 7). Congress proceedings and dissertations also provide much information in addition to that contained in standards, books and journals, but are more difficult to trace.

Project documentation is important because the history of the project is often a determining factor in its future.

Make sure that the analysis does not influence the result

It sometimes happens that the act of observation or measurement influences the result. If people either feel, or know, that they are under close scrutiny, their bodies will function differently (blood pressure, heart beat) and their behaviour will change (anticipation of desired outcome). These influences must be kept to a minimum by allowing those involved to become fully accustomed to the analysis method.

Selection of solutions phase

Enough must be known about a project before possible solutions can be devised. On the other hand, the search for a solution should initially not be limited by what might appear to be rigid project constraints. When making a purchase we will, in this phase, assess all available products. In design processes we will draw up an inventory of options. This is a point where ergonomics textbooks, software and other tools can make an important contribution in formulating ideas (see Chapter 7).

Realize that textbooks, software and other tools do not provide a complete answer

Textbooks, software (Figure 6.2) and other tools such as templates (Figure 6.3) and mannequins (Figure 6.4) never give a complete answer to questions such as how to design a control panel, a room or a product.

Allow users to work with a prototype

Many suggestions for improvement are made by users who are allowed to see a prototype and are given the opportunity to use it. This will increase the acceptance of the system at subsequent stages and ensure the commitment of the users to the system. Mock-ups can be used for workplace layouts, that is, full-scale models of the design. The material from which these mock-ups are made must allow rapid modifications; for example, wood or cardboard are quite suitable.

Figure 6.2 Example of a software tool to analyze human lifting.

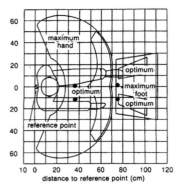

Figure 6.3 Example of a template which provides help in assessing drawings (view from above).

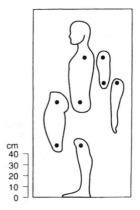

Figure 6.4 Example of a mannequin to illustrate the possibilities and limitations of the human body.

Computer simulations can also be used to predict reactions, and are most useful if they can later be used for training.

Remember the indirect users

Users other than direct users will also need to be able to deal with the new situation. Two such important target groups are cleaners and maintenance workers. Some designers make the mistake of assuming that cleaners and maintenance workers do not make full-time use of the design. However, this should not be taken to mean that sub-optimal solutions are acceptable. Indeed, the working day of a cleaner or maintenance worker might otherwise always consist of a succession of unsatisfactory situations.

Implementation phase

The next phase is to put in place the product or system which is selected as the best alternative. The following aspects play a role here:

- installation and testing;
- workplace design;
- training and support to users;
- organizational changes;
- acceptance of the new product or system.

Each of these aspects is dealt with below.

Select an implementation strategy beforehand

There are various ways of achieving a smooth transition from an old product or system to a new one. The choice depends on the type of project.

- From scratch: there is no old system. Staff recruitment requires special attention (group processes, training).
- Direct transition: at a given moment the old product or system is completely replaced by a new one. Both the organization and the technology change simultaneously. The output immediately following transition will be lower.
- Parallel application: the old and new systems are used side-by-side. The users must be well informed about the advantages of the new system, otherwise they will tend to use the old system too much since, after all, they are familiar with it.
- Phased introduction: successive phases of the new product or system are brought into use gradually. Acceptance can be more difficult because the advantage of the new solution is not imme-diately apparent to the users. Providing information is then very important. A particular form of this method is the introduction of a new system on a departmental basis.
- Occasional trial runs: this method is difficult to assess because the extra attention which the new system receives influences its use, just as in the case of the analysis.

Tests should be realistic

When testing the system or products, the real situation should be reproduced as closely as possible. This means, for example, that an automated database must react in the dialogue just as if it were completely filled with data. Involving the users in the tests can also double as training.

Train all users

Everyone who is involved with the system must receive some form of training. The training does not necessarily have to be the same for everyone. It is conceivable that key personnel could be given extensive training, after which they in turn can train their colleagues.

Support the implementation by providing good manuals

Good written explanations are an essential part of the implementation of the system. After all, users cannot be expected to remember all the finer points of a verbal explanation. The written documentation cannot, however, replace the verbal explanation. What is more, the objections and requirements given in Chapter 3, such as loss and limited accessibility of documents, also apply. Manuals act rather as a form of memory support.

The documentation must be suited to the desired objective. If the manual is intended to be used merely for reference, there is no point in giving a chronological process description. However, many users need information when something goes wrong, so that a manual must anticipate this. Many manuals describe operator actions, followed by their effect, whereas the user needs an overview of the effects, followed by the actions required. Manuals are definitely not a suitable means of compensating for compromises in the design.

Give the users a role in organizational changes

The tasks required to move from the old to the new organization must be properly assigned: responsibilities for specific aspects of the transition must be clear. These responsibilities are usually given to technical designers, instructors or ergonomists. It is also recommended to allocate roles to potential users, as this increases their commitment.

Convince the users of the improvements

If a project has led to a number of ergonomic improvements, it is crucial that the users be convinced of this. Some improvements have an effect only in the long term, such as the disappearance of aches and pains. Any disadvantages must also be pointed out. Indeed, there is no sense in pointing out the advantages while glossing over the disadvantages. Management must also be convinced of the improvements offered by the new product or system.

Evaluation phase

In time, it may appear, even in cases of careful selection or design, that all is not optimal. This will usually concern details where there is room for minor adjustments. Although many problems may appear to be solved in practice, the project still needs to be evaluated systematically after its introduction. It is necessary to assess whether the outcome meets the initial objectives, or to put it another way, whether the end result satisfies the initial objectives.

Keep to the same techniques of data collection

Whenever there is a redesign or readjustment, the result must afterwards be compared with the old situation. For this, it is advisable to use the same techniques as in the initial data collection, so that a proper comparison is possible. This means in the first instance a re-examination of the formal description of the activities on paper, and subsequently forming a clear picture of the situation by means of observations, interviews, etc.

Allow teething problems to sort themselves out

It is important not to carry out an evaluation too soon after implementation. Initially there will inevitably be teething problems which could lead to incorrect assessment of a new product or production system. This is also true of the attitude of users towards the new situation. If the modification consists of automating a system, they will initially find it interesting to work with the new product. In the course of time, the novelty will wear off and working with the new system may sometimes appear more monotonous than working with the old one.

Beware of habit formation

People are adaptable, and this can be an advantage. However, their ability to adapt may distort the result of an evaluation because they may have adapted to an undesirable situation. We must, therefore, carefully examine whether the use of the new product or system takes place in the correct i.e., intended, manner.

Checklist

A checklist can be used in different phases of an ergonomic approach, for example:

- to avoid overlooking any aspects;
- to detect where problems might develop;
- to measure the effect of the implementation;
- to obtain ideas for alternative solutions.

Prepare your own checklist, based on that of others

In practice an ergonomist will use an existing checklist as a basis for developing his own specific list. A checklist for an office environment will look very different from a checklist for the steel industry, and will emphasize entirely different topics.

The checklist should sometimes be combined with a scoring list so as to be able to carry out a reasonably objective ergonomic assessment of situations and a comparison thereof. The checklist presented is aimed only at the main points and allows only 'yes' and 'no' answers. The topics for consideration are based on the recommendations in the previous chapters.

SUMMARY CHECKLIST

Project factors

1　Are users involved in the project?
2　Are ergonomic requirements introduced as early as possible in the project?
3　Are conventional methods used for project management?
4　Is the planning flexible?
5　Is it understood that co-operation implies responsibility?

Initiative phase

6　Has who is involved in the project been defined?
7　Is the project supported by those who are involved?
8　Do team members agree on the code of conduct?
9　Are false expectations on the part of the user avoided?
10　Are the limits of the project stated?
11　Is the course of the project described at the onset?

Problem identification phase

12　Is the data processing established at the start?
13　Is more than one analysis technique selected?
14　Are existing documents surveyed at the start?
15　Is it assured that the analysis does not influence the result?

Selection of solutions

16　Is it understood that textbooks, software and other tools do not provide the complete answer?
17　Are users allowed to work with a prototype?
18　Are indirect users taken into account?

Implementation phase

19　Is an implementation strategy selected beforehand?
20　Are tests realistic?
21　Are all users trained?
22　Is the implementation supported by the provision of good manuals?
23　Do users have a role in organizational changes?

24   Are users convinced of the improvements?

Evaluation phase

25   Are the same techniques of data collection maintained?
26   Can teething problems sort themselves out?
27   Is there awareness for the possibility of habit formation?

Factors related to work organization, jobs and tasks

Tasks

28   Are tasks described in a neutral way?
29   Has a conscious decision been made about allocating tasks to a person, or to a machine?

Jobs

30   Does the job consist of more than one task?
31   Do those involved contribute to problem-solving?
32   Is the cycle time longer than one and a half minutes?
33   Is there alternation between easy and difficult tasks?
34   Can those involved decide independently on how the tasks are carried out?
35   Are there adequate possibilities for contacts with others?
36   Is the information provided sufficient to control the task?

Work organization

Flexible forms of organizations

37   Are hierarchical work organizations replaced by more flexible structures?
38   Are housing conditions and working times flexible?

Autonomous groups

39   Is the assignment to the team clear?
40   Does the team consist of 7–12 members?

Coaching management styles

41   Is the role of the manager more coach than boss?

Biomechanical, physiological and anthropometric factors

42   Are the joints in a neutral position?
43   Is the work held close to the body?
44   Are forward-bending postures avoided?
45   Are twisted trunk postures avoided?
46   Are sudden movements and forces avoided?
47   Is there a variation in postures and movements?
48   Is the duration of any continuous muscular effort limited?
49   Is muscle exhaustion avoided?
50   Are the breaks sufficiently short to allow them to be spread over the duration of the task?
51   Is the energy consumption for each task limited?
52   Is rest taken after heavy work?
53   Has account been taken of differences in body sizes?
54   Have the right anthropometric tables been used for specific populations?

Factors related to posture

55   Has a basic posture been selected that fits the job?

Sitting

56   Is sitting alternated with standing and walking?
57   Are the height of the seat and backrest of the chair adjustable?
58   Is the number of adjustment possibilities limited?
59   Have good seating instructions been provided?
60   Are the specific chair characteristics dependent on the task?
61   Is the work height dependent on the task?
62   Do the heights of the work surface, seat and feet correspond?
63   Is a footrest used where the work height is fixed?
64   Are excessive reaches avoided?
65   Is there a sloping work surface for reading tasks?
66   Is there enough legroom?

Standing

67   Is standing alternated with sitting and walking?
68   Is the work height dependent on the task?
69   Is the height of the work table adjustable?
70   Has the use of platforms been avoided?

71   Is there enough room for the legs and feet?
72   Are excessive reaches avoided?
73   Is there a sloping work surface for reading tasks?

Change of posture

74   Has an effort been made to provide a varied task package?
75   Have combined sit-stand workplaces been introduced?
76   Are sitting postures alternated?
77   Is a pedestal stool used once in a while in standing work?

Hand and arm postures

78   Has the right model of equipment been chosen?
79   Is the tool curved instead of the wrist being bent?
80   Are hand-held tools not too heavy?
81   Are tools well maintained?
82   Has attention been paid to the shape of handgrips?
83   Has work above shoulder level been avoided?
84   Has work with the hands behind the body been avoided?

Factors related to movement

Lifting

85   Have tasks involving manual displacement of loads been limited?
86   Have optimum lifting conditions been achieved?
87   Has care been taken that any one person always lifts less, and preferably much less, than 23 kg?
88   Have lifting situations been assessed using the NIOSH method?
89   Are the weights to be lifted not too light?
90   Are the workplaces suited to lifting activities?
91   Are handgrips fitted to the loads to be lifted?
92   Does the load have a favourable shape?
93   Have good lifting techniques been used?
94   Is more than one person involved in heavy lifting?
95   Are lifting accessories used?

Carrying

96   Is the weight of the load limited?

97   Is the load held as close to the body as possible?
98   Are good handgrips fitted?
99   Is the vertical dimension of the load limited?
100 Is carrying with one hand avoided?
101 Are transport accessories being used?

Pulling and pushing

102 Are pulling and pushing forces limited?
103 Is the body weight used during pulling and pushing?
104 Are trolleys fitted with handgrips?
105 Do trolleys have two swivel wheels?
106 Are the floors hardened and even?

Environmental factors

Noise

107 Is the noise level below 80 decibel?
108 Is annoyance due to noise avoided?
109 Are rooms perhaps too quiet?
110 Has a low-noise working method been chosen?
111 Are quiet machines used?
112 Are machines well-maintained?
113 Are noisy machines enclosed?
114 Is noisy work separated from quiet work?
115 Is there an adequate distance to the source of noise?
116 Is the ceiling used for noise absorption?
117 Are acoustic screens used?
118 Are hearing conservation measures suited both to the noise and to the user?

Vibration

119 Is uncomfortable body vibration avoided?
120 Is vibration 'white finger' due to hand-arm vibration avoided?
121 Are shocks and jolts prevented?
122 Is vibration tackled at source?
123 Are machines regularly maintained?
124 Is the transmission of vibration prevented?
125 Are measures at the individual level applied only as a last resort?

Illumination

126 Is the light intensity for orientation tasks in the range 10–200 lux?

127 Is the light intensity for normal activities in the range 200–800 lux?

128 Is the light intensity for special applications in the range 800–3000 lux?

129 Are large brightness differences in the visual field avoided?

130 Are the brightness differences between task area, close surroundings and wider surroundings limited?

131 Is the information easily legible?

132 Is ambient lighting combined with localized lighting?

133 Is daylight also used for ambient lighting?

134 Are light sources properly screened?

135 Are reflections and shadows prevented?

136 Is diffuse light used?

137 Is flicker from fluorescent tubes avoided?

Climate

138 Are people able to control the climate themselves?

·139 Is the air temperature suited to the physical demands of the task?

140 Is the air prevented from becoming either too dry or too humid?

141 Are hot or cold radiating surfaces avoided?

142 Are draughts prevented?

143 Is the climate possibly too hot or too cold?

144 Are materials which have to be touched neither too cold nor too hot?

145 Are equally heavy tasks grouped in the same room?

146 Are the physical demands of the task adjusted to the external climate?

147 Is the air velocity optimal?

148 Are undesirable hot and cold radiation prevented?

149 Is the time spent in hot or cold environments limited?

150 Is special clothing used when spending long periods in hot or cold environments?

Chemical substances

151 Is the concentration of chemical substances in air subject to limits (TLVs or other)?

152   Are carcinogenic substances avoided?
153   Are peak loadings prevented?
154   Is exposure to mixtures of substances prevented?
155   Is the concentration of chemical substances as far below the TLVs as possible?
156   Does the labelling on packages of chemicals provide information on the nature of any hazard due to the contents?
157   Can the source be removed?
158   Can releases from the source be reduced?
159   Can the source be isolated?
160   Are chemical substances extracted directly at the source?
161   Is the air exhaust system efficient?
162   Has attention been paid to climate at workplaces where exhaust and ventilation are used?
163   Are sufficient air changes provided?
164   Are organizational measures possible?
165   Is personal protective equipment available?
166   Are dust masks used only as protection against coarse dust?
167   Are protective clothing and gloves available if necessary?
168   Is attention paid to personal hygiene?

Factors related to information and operation

The user

169   Is the user-population defined as detailed as possible?
170   Are mental models of the users taken into account?

Information

Visual information

171   Have texts with only capitals been avoided?
172   Has justifying text by means of blank spaces been avoided?
173   Have familiar typefaces been chosen?
174   Has confusion between characters been avoided?
175   Has the correct character size been chosen?
176   Are longer lines more widely spaced?
177   Is the contrast good?
178   Are the diagrams easily understood?
179   Have pictograms been properly used?
180   Has an appropriate method of displaying information been selected?

Hearing

181 Are sounds reserved for warning signals?
182 Has the correct pitch been chosen?
183 Is synthesized speech adjustable?

Other senses

184 Are taste, smell and temperature restricted to warning signals?
185 Is the sense of touch used for feedback from controls?
186 Are different senses used for simultaneous information?

Controls

Distinguishing between controls

187 Are differences between controls distinguishable by touch?
188 Is the location consistent and has sufficient spacing been provided?
189 Is unintentional operation avoided?
190 Are controls well within reach?
191 Are labels or symbols properly used?
192 Is the use of colour limited?

Types of control

193 Has the QWERTY layout been selected for the keyboard?
194 Has a logical layout been chosen for the numerical keypad?
195 Is the number of function keys limited?
196 Is the type of cursor control suited to the task?
197 Is the mouse used not too frequently?
198 Are touch screens used to facilitate operation by inexperienced users?
199 Are pedals only used where the use of the hands is inconvenient?
200 Are remote controls used to give the user more freedom?

Relationship between information and operation

Expectation

201 Is the direction of movement consistent with expectation?
202 Is the objective clear from the position of the controls?
203 Is dual control used only where the consequences can be serious?

User-friendliness

204 Is the dialogue suitable for the task?
205 Is the dialogue self-descriptive?
206 Is the dialogue controllable?
207 Does the dialogue conform to the expectations on the part of the user?
208 Is the dialogue error-tolerant?
209 Is the dialogue suitable for individualization?
210 Is the dialogue suitable for learning?

Different forms of dialogue

211 Have menus been used for users with little knowledge and experience?
212 Are the limitations of an input form known?
213 Has command language been restricted to experienced users?
214 Is direct manipulation consistent?
215 Have the disadvantages of natural language been recognized?

Help

216 Is the type of help suited to the level of the user?

Concluding factor

217 Would you wish to carry out the task yourself?

7 Sources of additional information

Compiled by E.D. Megaw

This chapter provides information to readers wishing to learn more about ergonomics. We first cite some of the general and more specialist ergonomics sources which were used in compiling the present book. This is followed by a listing of the most relevant ergonomics journals. The chapter concludes with details of some key websites.

Bibliography

The background to the guidelines and advice given in this book can be found in the literature listed below. The more general sources appear first, followed by the more specialist sources grouped according to the chapters of this book. If you wish to get details of other books on ergonomics it is suggested you access the Amazon website (www.amazon.com).

General references

Bridger, R.S., 1995, *Introduction to Ergonomics* (New York: McGraw-Hill) ISBN 0 07 007741 X

Helander, M., 1995, *A Guide to the Ergonomics of Manufacturing* (London: Taylor & Francis) ISBN 07484 0122 9

Karwowski, W. and Marras, W.S. (eds), 1999, *The Occupational Ergonomics Handbook* (Boca Raton: CRC Press) ISBN 0 849 32641 9

Konz, S. and Johnson, S., 2000, *Work Design: Industrial Ergonomics* (5th edition) (Scottsdale, Arizona: Holcomb Hathaway) ISBN 0 890871 07 9

Kroemer, K.H.E. and Grandjean, E. (eds), 1997, *Fitting the Task to the Human: a Textbook of Occupational Ergonomics* (5th edition) (London: Taylor & Francis) ISBN 0 7484 0665 4

Norman, D.A., 1998, *The Psychology of Everyday Things* (2nd edition) (London: MIT Press) ISBN 0 262 64037 6

Oborne, D.J., 1995, *Ergonomics at Work* (3rd edition) (Chichester: Wiley) ISBN 0 471 95235 4

Salvendy, G. (ed.), 1997, *Handbook of Human Factors and Ergonomics* (2nd edition) (New York: John Wiley) ISBN 0 471 11690 4

Sanders, M.S. and McCormick, E.J., 1992, *Human Factors in Engineering and Design* (7th edition) (New York: McGraw-Hill) ISBN 0 07 112826 3

Wickens, C.D., Gordon, S.E. and Liu, Y., 1998, *An Introduction to Human Factors Engineering* (New York: Longman) ISBN 0 321 01229 1

Woodson, W.E., Tillman, B. and Tillman, P., 1992, *Human Factors Design Handbook* (2nd edition) (New York: McGraw-Hill) ISBN 0 07 071768 0

References on posture and movement

Cacha, C.A., 1999, *Ergonomics and Safety in Hand Tool Design* (Boca Raton, Florida: Lewis Publishers) ISBN 1 56670 308 5.

Chaffin, D.B., Martin, B.J. and Andersson, G.B.J., 1999, *Occupational Biomechanics* (3rd edition) (New York: John Wiley) ISBN 0 471 24697 2

Corlett, E.N. and Clark, T.S., 1995, *The Ergonomics of Workspaces and Machines: a Design Manual* (2nd edition) (London: Taylor & Francis) ISBN 0 7484 0320 5

Pheasant, S., 1996, *Bodyspace: Anthropometry, Ergonomics and Design* (2nd edition) (London: Taylor & Francis) ISBN 0 7484 0326 4

Kumar, S. (ed.), 1999, *Biomechanics in Ergonomics* (London: Taylor & Francis) 0 7484 0704 9

Mayer, T.J., Gatchel, R.J. and Polati, P.B. (eds), 2000, *Occupational Musculoskeletal Disorders Function, Outcomes, and Evidence* (Philadelphia: Lippincott Williams & Wilkins) ISBN 0 7817 1735 3

Mital, A., Nicholson, A.S. and Ayoub, M.M., 1997, *A Guide to Manual Materials Handling* (2nd edition) (London: Taylor & Francis) ISBN 0 7484 0728 6

Nordin, M., Andersson, G.B.J. and Pope, M.H., 1997, *Musculoskeletal Disorders in the Workplace: Principles and Practice* (St Louis, Missouri: Mosby) ISBN 0 8016 7984 2

Putz-Anderson, V. (ed.), 1988, *Cumulative Trauma Disorders. A Manual for Musculoskeletal Diseases of the Upper Limbs* (London: Taylor & Francis) ISBN 0 85066 405 5

References on information and operation

Dix, A.J., Finlay, J.E., Abowd, G.D. and Beale, R., 1998, *Human–Computer Interaction* (2nd edition) (London: Prentice Hall Europe) ISBN 0 13 239864 8

Downton, A. (ed.), 1993, *Engineering the Human–Computer Interface* (student edition) (London: McGraw-Hill) ISBN 0 07 707727 X

Shneiderman, B., 1998, *Designing the User Interface* (3rd edition) (Reading, Massachusetts: Addison-Wesley) ISBN 0 201 69497 2

Pearrow, M., 2000, *Website usability handbook* (Hingham: Charles River Media) ISBN 1 5845 0026 3

Zwaga, H.J.G., Boersema, T. and Hoonhout, H.C.M. (eds), 1999, *Visual Information for Everyday Use: Design and Research Perspectives* (London: Taylor & Francis) ISBN 0 7484 0670 0

References on environmental factors

Chartered Institution of Building Services Engineers, 1994, *Code for Interior Lighting 1994* (London: CIBSE) ISBN 0 900953 64 0

Griffin, M.J., 1990, *Handbook of Human Vibration* (London: Academic Press) ISBN 0 12 303040 4

Kleeman, W.B., 1991, *Interior Design of the Electronic Office. The Comfort and Productivity Payoff* (New York: Van Nostrand Reinhold) ISBN 0 442 00613 6

Kryter, K.D., 1994, *The Handbook of Hearing and the Effects of Noise: Physiology, Psychology, and Public Health* (San Diego, California: Academic Press) ISBN 0 12 427455 2

Parsons, K.C., 1993, *Human Thermal Environments* (London: Taylor & Francis) ISBN 0 7484 0041 9

Rostron, J. (ed.), 1997, *Sick Building Syndrome: Concepts, Issues and Practice* (London: Spon Press) ISBN 0 419 21530 1

References on work organization

Brosnan, M.J., 1998, *Technophobia: The Psychological Impact of Information Technology* (London: Routledge) ISBN 0 415 13597 4

Chmiel, N., 1998, *Jobs, Technology and People* (London: Routledge) ISBN 0 415 15817 6

Eason, K., 1988, *Information Technology and Organisational Change* (London: Taylor & Francis) ISBN 0 85066 388 1

Hackman, J.R. and Oldham, G.R., 1980, *Work Redesign* (London: Addison-Wesley) ISBN 0 201 02779 8

Monk, T.H. and Folkard, S., 1992, *Making Shift Work Tolerable* (London: Taylor & Francis) ISBN 0 85066 822 0

Parker, S. and Wall, T., 1998, *Job and Work Design: Organizing Work to Promote Well-Being and Effectiveness* (Thousand Oaks, California: SAGE Publications) ISBN 0 7619 0420 4

Pasmore, W.A., 1988, *Designing Effective Organizations: The Sociotechnical Systems Perspective* (New York: John Wiley) ISBN 0 471 88785 4

136 Ergonomics for Beginners

References on the ergonomic approach

Annett, J. and Stanton, N.A. (eds), 2000, *Task Analysis* (London: Taylor & Francis) ISBN 0 7484 0906 8.

Jordan, P.W., 1998, *An Introduction to Usability* (London: Taylor & Francis) ISBN 0 7484 0726 6

Jordan, P.W., Thomas, B., Weerdmeester, B.A. and McClelland, I.L. (eds), 1996, *Usability Evaluation in Industry* (London: Taylor & Francis) ISBN 0 7484 0460 0

Kirwan, B. and Ainsworth, L.K. (eds), 1992, *A Guide to Task Analysis* (London: Taylor & Francis) ISBN 0 7484 0058 3

Rubin, J., 1994, *Handbook of Usability Testing* (New York: John Wiley) ISBN 0 47159403 2

Stanton, N.A. and Young, M.S., 1999, *A Guide to Methodology in Ergonomics: Designing for Human Use* (London: Taylor & Francis) ISBN 0 7484 0703 0

Wilson, J.R. and Corlett, E.N. (eds), 1995, *Evaluation of Human Work: a Practical Ergonomics Methodology* (2nd edition) (London: Taylor & Francis) ISBN 0 7484 0084 2

Scientific and professional journals

Applied Ergonomics (Amsterdam: Elsevier)

Behaviour and Information Technology (London: Taylor & Francis)

Cognition, Technology & Work (Berlin: Springer)

Ergonomics (London: Taylor & Francis)

Ergonomics Abstracts (London: Taylor & Francis)

Ergonomics in Design (Santa Monica, California: Human Factors and Ergonomics Society)

Human–Computer Interaction (Mahwah, New Jersey: Lawrence Erlbaum Associates)

Human Factors (Santa Monica, California: Human Factors and Ergonomics Society)

Human Factors and Ergonomics in Manufacturing (New York: John Wiley)

Interacting with Computers (Amsterdam: Elsevier)

International Journal of Cognitive Ergonomics (Mahwah, New Jersey: Lawrence Erlbaum Associates)

International Journal of Human–Computer Interaction (Mahwah, New Jersey: Lawrence Erlbaum Associates)

International Journal of Human–Computer Studies (London: Academic)

International Journal of Industrial Ergonomics (Amsterdam: Elsevier)

International Journal of Occupational Safety and Ergonomics (Warsaw: Central Institute for Labour Protection)

Japanese Journal of Ergonomics (Tokyo: Business Center for Academic Societies Japan)

Occupational Ergonomics (Amsterdam: IOS)

Scandinavian Journal of Work, Environment & Health (Helsinki: Finnish
   Institute of Occupational Health)
Theoretical Issues in Ergonomics Science: TIES (London: Taylor & Francis)
Tijdschrift voor Ergonomie (Nieuwegein: Nederlandse Vereniging voor
   Ergonomie)
Travail Humain (Paris: Presses Universitaires de France)
Work and Stress (London: Taylor & Francis)
Zeitschrift für Arbeitswissenschaft (Cologne: Otto Schmidt)

Useful websites

The websites listed below are just a few of the available sites devoted
to ergonomics. However, the links found in these sites will allow you
to gain access to most of the other available sites.

International Ergonomics Association (IEA): the International
Ergonomics Association (IEA) comprises approximately 40 member
societies representing about 15 000 ergonomists worldwide.
   http://www.iea.cc

Ergonomics Information Analysis Centre
   http://www.bham.ac.uk/manmecheng/ieg/eiac

Ergonomics Society, UK
   http://www.ergonomics.org.uk

ErgoWeb
   http://www.ergoweb.com

Human Factors and Ergonomics Society, USA
   http://www.hfes.org

Human Systems Information Analysis Center
   http://iac.dtic.mil/hsiac

Usernomics
   http://www.usernomics.com

Authors' website
   http://www.ergonomicsforbeginners.com

Index

noise (*cont'd*)
    reduction 76, 77
noisy machines 77
numerical keypad 58

optimum lifting conditions 33

participatory ergonomics 112
peak exposures 93
pedals 60
pedestal stool 13, 24
performance 1
personal hygiene 98
personal protective equipment 97
physiology 9
pictograms 50
platforms 21
posture 5, 12
prefix 64
preventing vibration 82
project management 112, 113
prolonged postures 7, 22
prolonged sitting 14
protective equipment 98
prototype 118
pulling and pushing 38
    force 39
    posture 39

quiet machines 77
quiet working method 77
QWERTY keyboard 58

radiating surfaces 89
reach 18, 56
reaches 22
recovery 9
reflectance 87
reflections 87
remote controls 61
Repetitive Strain Injuries or RSI
    25, 60

repetitive movements 7

safety 1
screwdrivers 25
seating instructions 15
shape of handgrips 27
shapes 55
shocks and jolts 82
shoulder complaints 14, 20, 25
sit-stand workstation(s) 13, 23
sitting 13
    workstation 13
sloping work surface 18, 19, 22
smell, taste and touch 53
software tool 118
sound-insulating enclosure 77
space 97
standards 3
standing 20
    workstation 13
steering wheels 54
sudden movements 7
suffix 64
swivel wheels 40
symbols 57
synthesized speech 53

tasks 102
template 119
tennis elbow 24
thermal comfort 88
threshold limit values 92
tilt 15
TLV 92
tool 25
touch-screens 60
transport accessories 38
twist 15, 18
twisted trunk 6

user-friendliness 64
user-interface 44

142   *Index*